IT and microelectronics

Peter Bishop

Series editor: Graham Hill

Contents

Introducing IT:
Forecasting the weather 3
1. Systems 4
2. Inputs and outputs 6
3. Sensors 8
4. Monitoring systems 10
5. Controlling systems 12
6. Automatic control 14
7. Switches and relays 16
8. Analogue and digital signals 18
9. Logic gates and circuits 20
10. Storing information 24
11. Communicating information 26
12. Displaying information 28

Bar code scanners are an important part of the information technology (IT) in most supermarkets. They use complicated microelectronic components

Activities

1. Rolling race 30
2. Choosing an athletics team 31
3. Radar and echo sounders 32
4. Monitoring a chemical reaction 33
5. Pulse rates 34
6. Electrocardiographs 35
7. Rapid reactions? 36
8. Counting made easy 38
9. Keeping warm 40
10. Marks and spaces 42
11. Keeping tropical fish 44

Glossary 47
Equipment 47

The robots on this factory production line can do repetitive jobs which were once done by workers. Robots are programmed using computers

Introduction

Science Scene is a series of books that will help your studies in Key Stage 3 of the National Curriculum.

Each book in the series looks at one of the attainment targets in science. This book – *IT and microelectronics* – covers the scientific aspects of information technology including microelectronics. It introduces the key ideas which enable you to understand the topic. It also lets you put these key ideas into practice in a range of activities.

There are two parts to the book:

- The first part introduces the key ideas you need. Some of these, like *input* and *output*, may be new to you. The best way to get used to them is to think of the many examples which you will find in everyday life. These ideas are useful in science, and in many other activities.

- The second part of the book contains a number of activities. These put the ideas from the first part into practice. Some of them involve the use of computers. Others enable you to use electronics kits. Although the activities are simple enough for you to do, they are similar to the way experiments are done in research laboratories.

At the end of the book is a glossary, which explains the technical words used in the book. There is also a list of the equipment you need for the activities.

If you work through the book, you will cover all the work that is needed for Microelectronics and IT at Key Stage 3. You will have covered both the theory and the practical work.

We hope you will enjoy working through *IT and microelectronics*, and that you will learn a lot which is new.

Peter Bishop and Graham Hill (1990)

Introducing IT: Forecasting the weather

This television weather forecaster tells us the sort of weather we can expect

Weather forecasts, like the ones on television, are very important. We all want to know whether tomorrow will be hot and sunny, or cold and wet, because then we can make plans. For some people, weather forecasts are vital. Farmers, sailors and pilots depend on the weather because it affects their work. Weather forecasts can warn all of us about problems like storms and hurricanes. Sometimes these warnings can prevent accidents.

Weather forecasting is difficult because the weather is very complex. It varies from place to place, and it is caused by many different factors. Predicting the weather is a science. Weather forecasters have to observe, take measurements, record information and then process it, just as you do in science investigations. Today computers help to make weather forecasting faster and more accurate. The computers are used to record, store and process information.

Weather stations, balloons, ships, satellites and aircraft take lots of measurements of temperature, pressure, humidity, wind speed, etc.. They feed their results into some of the largest and most powerful computers in the world. The computers process this information and give an idea of how the Earth's atmosphere will change over the next few days. All the computers and electronic instruments are linked together on a worldwide network so they can share information. The weather forecasters use the results to make their own predictions.

In modern weather forecasting, computers, communication links and electronic equipment work together. This is one example of **information technology**, or **IT**. It shows how important IT is in science.

This book introduces you to some of the key ideas behind the way information technology and microelectronics are used in science and in everyday life. The ideas are quite simple and they are also very useful. They help to make complicated things easy to investigate. Once you have read about the key ideas, you can put them into practice by trying the activities in the second part of the book.

Computers can help us to make weather forecasts faster and more accurate. This can help us to prevent damage from storms and hurricanes like that in the photo above

Things to do

1 *Red sky at night, shepherd's delight.*
Red sky in the morning, shepherd's warning.

This old saying shows how people once predicted the weather.
  a) How do you think this saying first started?
  b) Some people still use sayings like this to predict the weather. How scientific do you think this type of weather forecasting is?
  c) Find out about 3 more sayings that predict sun, rain and snow. Are they accurate ways of predicting the weather? Prepare a short talk for your class, explaining how useful you think the sayings are.

2 Science books and magazines often have photos of computers being used in experiments. Try to find 2 or 3 photos, and make a poster of them. Beside each photo:
  a) describe the experiment that is taking place;
  b) say what the computer is used for;
  c) explain the advantages of using the computer in the experiment.

1 Systems

The human body and a car seem very different. Can you think of any similarities between them?

Human bodies and cars are different in lots of ways, but they are alike in some ways too. Both are made up of parts that work together.

A car has an engine, a battery, wheels, seats, and so on. Each part performs a task, but alone, it is not much use. When the parts work together, they form a car.

Your body is also made up of parts, such as your heart, lungs, stomach, skeleton, and so on. Each part has a job to do, but it cannot live on its own. Working together, they form your body.

Things, like cars and our bodies which contain parts that work together, are called **systems**.

> A **system** is a set of parts, working together to do something useful.

Systems can be **natural**, like your body, or **constructed**, like a car or a bicycle. They can be simple, like a pencil sharpener, or complicated, like a television set.

Thinking of things as systems is useful because it makes complicated things easier to understand. We can divide a whole system into smaller parts, and look at each of these separately.

Things to do

1 Look at the diagram of a torch below. It has only a few parts: bulb, reflector, case, batteries and switch. Each part has a task of its own. For example, the bulb produces the light.
  a) What do you think the other parts do?
  b) What would happen if one of the parts did not work?

3 The photos below show systems.
  a) Which systems in these photos are natural and which are constructed?
  b) Which systems are simple, and which are complicated?
  c) Which of these photos do not show systems? In each case, explain why you think it is not a system.

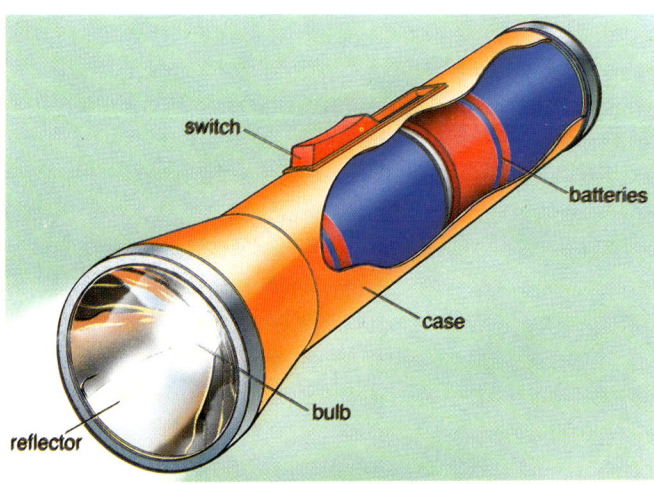

2 Look at the plant in the photo. Is it a simple natural system? Sketch the plant, and label the main parts. Next to each label, write down what the part does.

2 Inputs and outputs

What are the inputs into the systems shown in these five photos?

You don't always need to know how a system works. But it is usually important to know two things about a system: what goes into it, and what comes out of it.

Anything which goes into a system is called an **input**.

Food is an input for our bodies. Petrol is an input for a car. Inputs do not have to be substances, like food and fuel. The television signal is an input for a television set. Without it, the TV would not work. Information can also be an input. We take in information when we read a book, for example.

What are the outputs from the systems shown in these three photos?

Anything which comes out of a system is called an **output**.

Milk from a cow, and metal from a furnace are examples of outputs. Like inputs, outputs do not have to be substances. Picture and sound are the outputs from a TV set.

The output from one system can be the input into another system. For example, when you watch television, the output from the television becomes an input for your body.

IT AND MICROELECTRONICS

The New York skyline at night

The lights you can see in the photo are powered by electricity. Electricity is the input for the different systems which produce light. You turn on a light using a switch, which allows an electric current to flow into the system. The output of the system is light.

Things to do

1 Look at the drawing below and write down the different inputs and outputs that you can see.

2 The log fire in the photo on the right is a simple natural system with inputs and outputs.
   a) What are the inputs?
   b) What are the outputs?
   c) How are the outputs from the fire used as inputs by the other systems in this photo?

3 Work in a group with 2 or 3 others. Try to identify all the inputs and outputs of the telephone, the human body and the food processor below.

7

3 Sensors

Three different types of sensor

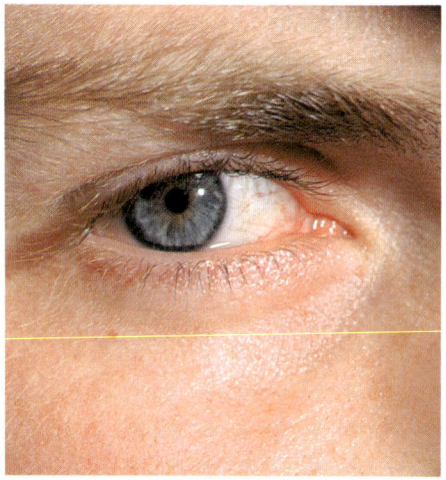 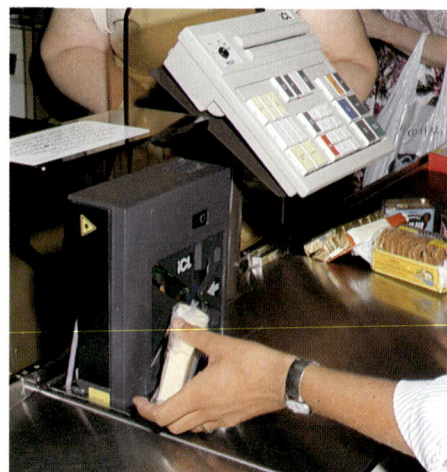

Some systems need to detect inputs and outputs. For this they use **sensors**. There are sensors in natural and constructed systems. Sensors detect signals like sound, light and heat. The photos above show various sensors. Can you name the sensor or sensors in each picture? What signal does each one detect?

There are many different types of sensors. In natural systems, complicated sensors like ears, eyes and noses, are very important to animals. They help them to stay alive, for example by helping them to find food, and to avoid danger. In constructed systems, there are also many different sensors. For example, the aerial on a radio is a sensor which detects radio waves.

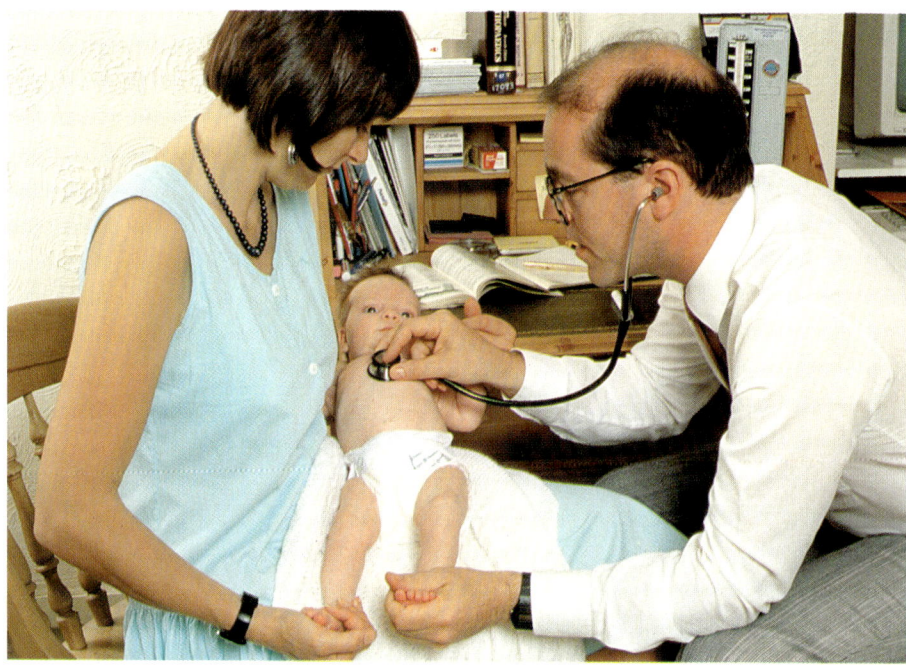

This doctor is using a sensor called a stethoscope. What signal does a stethoscope detect?

We use sensors in experiments. The sensors measure things like temperature and pressure. Many sensors are electronic. They have no moving parts and they convert the signals they receive into electric currents.

IT AND MICROELECTRONICS

Things to do

1 What sensors does this cat have? What is each sensor used for?

2 The smoke alarm below is a system. It has a sensor which detects smoke. When smoke is present, the alarm bell rings to warn people that there might be a fire.

a) What is the input and the output of this system?
b) What sensor in our bodies do we use to detect the output from a smoke alarm?
c) Draw a diagram to show the inputs and outputs for a smoke alarm warning someone of the danger of fire.
d) There are lots of examples of warning systems like this smoke alarm. Can you think of any others that work in a similar way? For each one, draw a diagram to show the inputs and outputs. (*Hint*: the output from the alarm does not have to be a noise.)

3 Look at the three photos below and then discuss these questions in groups of 3 or 4 people.

a) Does the Venus flytrap have any sensors? If so, what signals do the sensors respond to? How does the plant react?

b) What do the sensors in the metal detector respond to? Why is the detector useful?

c) What do you think this dog is looking for?

9

4 Monitoring systems

Look at the photograph of baby Ruby. Ruby was born too early and she is now in an incubator. Here she is kept warm and fed through tubes. Ruby's heartbeat, temperature and breathing must be checked at regular intervals. If there are any changes in these, the doctors and nurses must act quickly.

Years ago, it was difficult to carry out these checks all the time. Before electronic systems were used, babies born as early as Ruby had little chance of surviving. Today we have electronic equipment which takes measurements all the time. Babies like Ruby now have a good chance of survival.

Ruby was a premature baby. This means that she was born before she was fully developed. Her heartbeat, temperature and breathing are being measured all the time

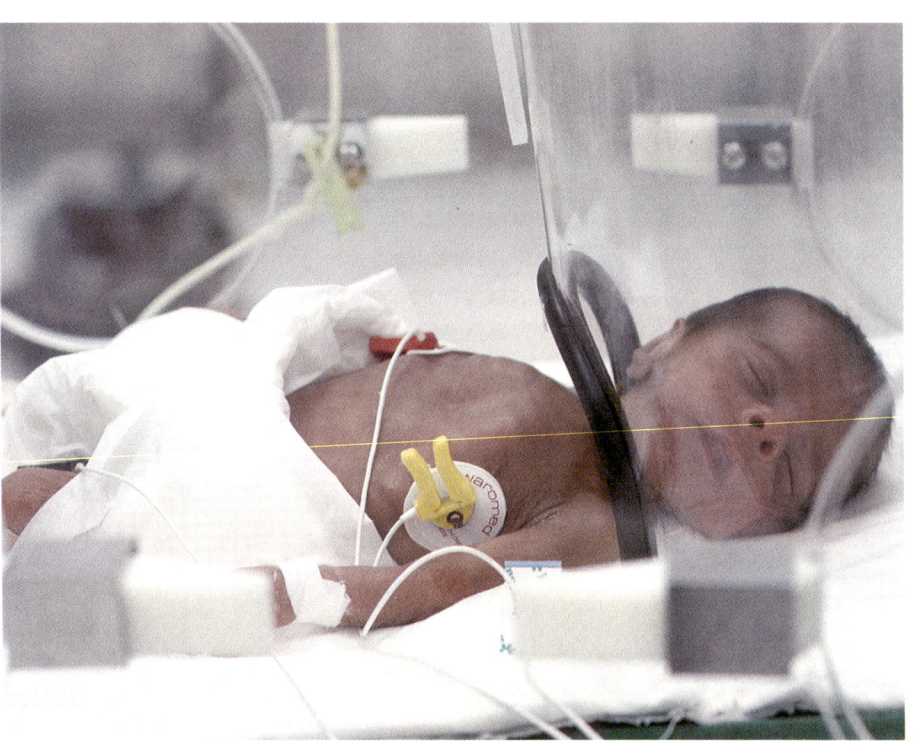

Taking measurements regularly is called **monitoring**.

Monitoring is important because it allows us to control systems. Systems like small babies, power stations, car engines, and ovens are just four examples. In every case, sensors are used to tell us about the system. In some systems, measurements are taken continuously. For example, in a power station, the current generated is measured and recorded all the time. In other systems, the sensor is used to detect a change. For example, the sensor in an oven detects when the temperature falls below the set temperature. If this happens, the heater is turned on, and the oven is reheated. Sensors like this one are often set so that they respond if the input goes above or below certain limits. Central heating systems and refrigerators work in the same way.

The control room at Calder Hall nuclear power station, part of the Sellafield plant in Cumbria. In the control room of a power station, the output from the generators is monitored all the time

IT AND MICROELECTRONICS

Monitoring is done by sensors. Many scientific experiments are monitored. They often have electronic sensors which monitor inputs or outputs, and feed the information into a computer for storage. A computer is needed because a sensor running all the time produces a great deal of information during the experiment.

The sensor monitoring this experiment is passing the information it gathers into a computer to be stored

Things to do

1 Look at the photographs below and to the right.
  a) What is being monitored (temperature, pressure, electric current, etc.) in each system?
  b) In each case, give a reason for taking the measurement(s) all the time.

2 A seismograph like the one below monitors vibrations in the earth, caused by earthquakes, volcanoes and explosions. There is a network of seismographs at research stations throughout the world.
  a) Why is it important to monitor vibrations in the earth all the time?
  b) What information can we get from the trace produced by a seismograph?

5 Controlling systems

How does sweating help to keep bodies cool?

Putting on the brakes is an important way to control a bicycle!

It is important to keep your body at a constant temperature. Chemical reactions take place when food is broken down. These reactions produce heat which keeps you warm. If you run around, or do exercises, your body gets too hot. You then begin to sweat, to cool down. Sweating is an important example of the **control of a system**.

> A **control** is something which makes a system work in a certain way.

Your body has lots of other natural controls. For example, when you run around you need more oxygen. The nerves which control your lungs make you breathe faster and take deeper breaths. Can you think of any others?

Constructed systems almost always have some form of control. For example, the fork-lift truck in the photo has brakes to control its speed, and a steering wheel to control its direction. When people design a system, the method of control that they design is very important.

How is this fork lift truck controlled? Think of the different forms of control that it must have

Almost all experiments are controlled in some way. For example, when we investigate how plants grow from seeds, we control the temperature of the seedlings and the amount of light and water they receive. Most chemical reactions are affected by temperature. For this reason, electronic systems are used to carefully monitor and control temperature in many industrial processes.

This scientist is using an electronic control panel to monitor a chemical reaction

Things to do

1 Make a list of all the controls you can think of in:
   a) an electric toaster;
   b) a personal stereo;
   c) an earthworm.

2 Most central heating systems in our homes have sensors and controls like the ones shown in the diagram below. The thermostat in a room controls the pump which sends hot water to the radiators. The temperature sensor in the boiler controls the flame or electric heater which warms the water.

Think about the way in which a central heating system works and try to explain how:
   a) the water pump is controlled by the room thermostat;
   b) the heater is controlled by the sensor in the boiler.

3 Chemical reactions take place in food during cooking. It is important to control these reactions, otherwise the food may be inedible. If it is cooked too much the food might be burnt, and if it is cooked too little it might be raw.

Imagine that you are preparing these foods for a meal. Describe from the start how you would control the systems that cook the foods to make sure that each one is done perfectly. (If you have never cooked any of these foods, then try to find out from someone who has.)
   a) A boiled egg.
   b) Fried rice.
   c) Bread.
   d) Christmas pudding.

How can you tell when each of these foods is ready?

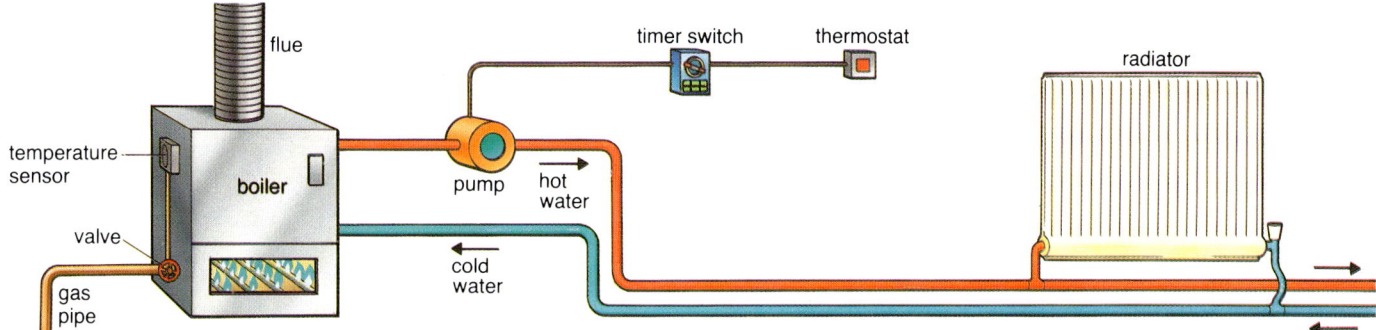

6 Automatic control

Traffic was once controlled by policemen. Now we use traffic lights

25 years ago, the traffic at busy junctions in cities was controlled by policemen. Would you have liked to do this job? Do you think it was dangerous? It was very difficult to co-ordinate all the junctions in a street so that traffic flowed smoothly.

Today, traffic junctions are controlled by lights which work automatically. No one has to press a switch to make them change colour. In most cities, all the traffic lights are controlled by computer. This central control helps to keep the flow of traffic steady. It can also be adjusted for different times of day, and to take account of accidents and roadworks.

Automatic control like this occurs when the control signals are produced within the system itself.

The robot trolley on the right moves goods in a factory without anyone driving it. It is controlled automatically

A simple example of automatic control is in a refrigerator (figure 1). It has a thermostat which monitors the temperature inside. If the temperature rises above a certain level, the pump is switched on. The pump sends cooling fluid around the pipes inside the refrigerator. This lowers the temperature in the refrigerator. When the temperature has dropped to the required level, the pump is switched off again. The temperature at which the pump switches on or off can be adjusted.

Figure 1 The automatic control system of a refrigerator

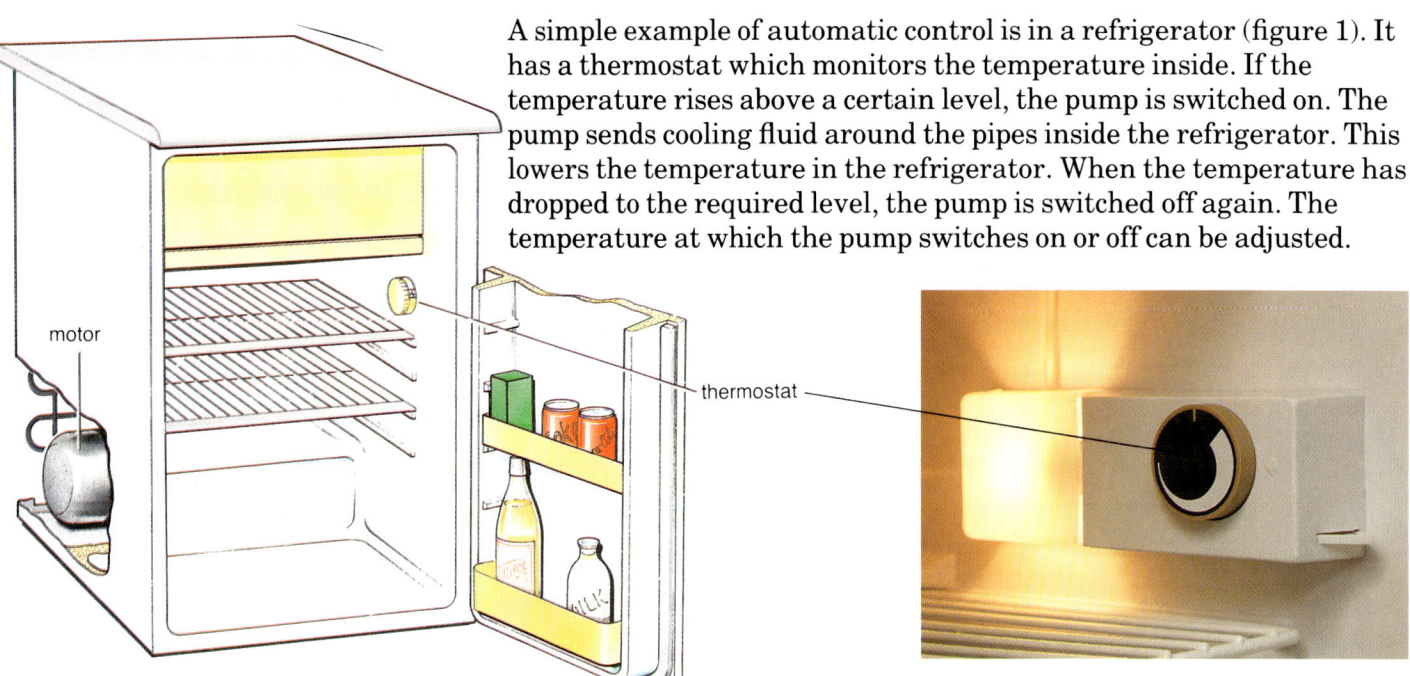

IT AND MICROELECTRONICS

We can use automatic control to prevent burglaries. If you leave your house empty at night, it is a good idea to leave lights and a radio on. The house will look occupied even though no one is at home, and so there is less chance of a break-in. You can use automatic systems to switch on radios, lights and even close curtains.

Most of the controls in plants and animals are automatic, or at least partly automatic. The control of your lungs is a good example of partly automatic control. You can deliberately take a breath, but if you forget about breathing, the automatic control takes over.

Constructed systems are making more and more use of automatic control. Many of the automatic controls make use of electronic devices or computers. Scientific experiments often have automatic controls. These controls might be used to keep things at a constant temperature. Automatic control is quicker, cheaper and usually safer than control by people (called **manual control**). For example, the safety systems in chemical plants are almost entirely automatic, because a person could not react quickly enough if something went wrong.

A helicopter sprays liquid on the wreckage of the nuclear reactor at Chernobyl in the USSR. The accident was caused by someone interfering with the automatic control of the reactor. Because of this, the system could not control itself when it went wrong

Things to do

1 A rule for the automatic control of a refrigerator can be written like this:

If the temperature is too high
   then switch the pump on.
   else switch the pump off.

Write similar rules for the following examples:
  a) the pump of the heating system in a swimming pool;
  b) an automatic door at a supermarket entrance, controlled by a sensor.

2 The photos below show a set of traffic lights on an isolated junction. The lights usually work in a fixed cycle, without any inputs. Make a list of 3 other automatic systems like this, where there are no inputs.

3 Make a list of 6 controls in your body which are partly automatic, like the control of your breathing.

4 Design a poster which would help 10-year-olds to understand the automatic control system of the oven in a gas cooker.

5 The 3 systems listed below are controlled automatically. For each one, draw a simple diagram showing how the automatic control works. Label the inputs, outputs and sensors. Briefly describe how the control system operates.
  a) A washing machine.
  b) The automatic pilot of an aircraft.
  c) A robot used to paint new cars on a factory assembly line.

7 Switches and relays

The simplest way to control a system is to use a **switch**. You put a switch in an electric circuit to control the current (figure 1).

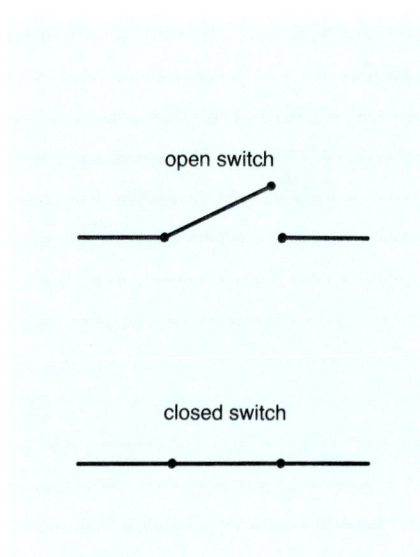

Figure 2 An open switch (off) and a closed switch (on)

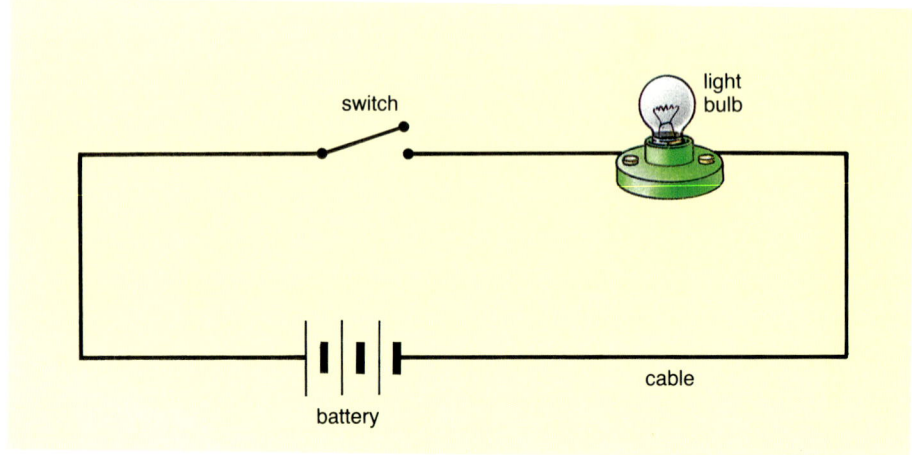

Figure 1 A switch in an electric light circuit

We say that a switch is open (off) when no current flows. The switch is closed (on) when the current flows. Figure 2 shows how you should draw switches in an electric circuit.

These photos show different switches in use. Photo-electric cells are used to time when the winner of a race crosses the finish line

This kettle has a built-in switch, and there is another one at the wall socket

You would find these switches in a car. What are they used for?

Switches can be used as sensors for different systems. For example, special types of switches can be operated by temperature. These are called **heat sensors** or **thermostats.** As the temperature changes, the thermostat will allow the current to start flowing, or it will stop it flowing. Other types of switch are operated by light. These are called **light sensors** or **photo-electric cells.** They produce an electrical signal when an object passes in front of them.

Some switches operate other switches. These other switches are called **relays** (figure 3). In the relay in figure 3, a switch is closed in the left hand circuit to make a small electric current flow. This current magnetises the iron bar which attracts the relay switch and makes it close. A large current then flows in the right hand circuit. In this way, a control system can operate with a *small* electric current, and use a relay to control a system which works with a *large* current. Power stations and electric locomotives are examples of systems which use relays.

An electric locomotive like this one has relay switches to control the large currents it uses

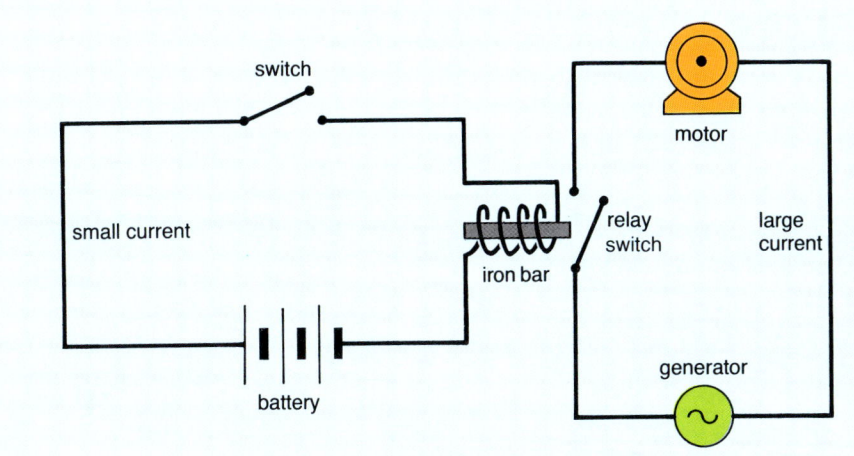

Figure 3 This is a relay switch controlling an electric motor

Things to do

1 Why do you think that relay switches are useful in power stations and electric trains?

2 In addition to switching currents on or off, switches can be used to select one of two circuits. These two-way switches are very useful. A two-way switch is shown in the figure below. Draw a diagram to show how two of these switches can be connected together to control a staircase light: one switch is at the bottom of the stairs, the other is at the top. If the light is on, changing one switch turns it off. If the light is off, changing one switch turns it on.

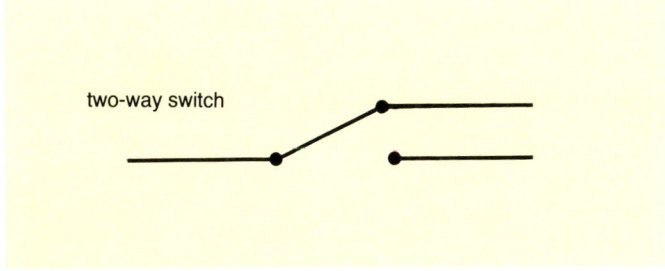

3 A thermostat (which senses temperature) can be used as a switch. It can allow current to flow in a circuit, or it can stop the current flowing.

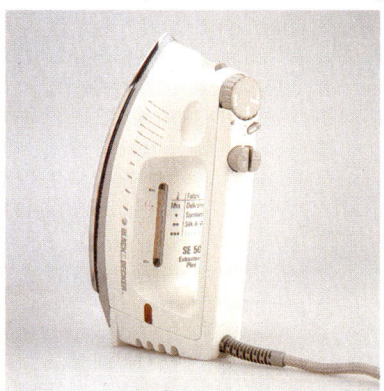

The temperature of this iron is controlled by a thermostat

a) Describe how a thermostat works.
b) You can use a thermostat to control an electric oven. Describe how the control system would work.

17

8 Analogue and digital signals

The needle on these bathroom scales has moved to indicate Tasneem's weight. The movement of the needle is an analogue signal

If you stand on scales like the ones shown in the picture, the needle moves around the scale. It stops when it shows your weight. The heavier you are, the further the needle moves. This is an **analogue signal**. The distance the needle moves is proportional to the quantity that it measures.

> An **analogue** signal is proportional to the quantity it measures. It can have a range of values.

When an object passes in front of a photo-electric cell, it blocks the light reaching the cell. The current (signal) produced by the photo-electric cell goes from a high value to nothing. The signal does not depend on the size of the object nor on its weight. All the signal does is show that an object has passed. This is an example of a **digital signal**.

A photo-electric cell counts the number of milk bottles on this factory production line. The movement of bottles past the cell produces a digital signal

> A **digital** signal is one which is either on or off. It cannot have a value in between.

A digital temperature sensor works in a similar way to the photo-electric cell. If the temperature is below a certain level, it gives no signal. When the temperature passes this level, the sensor sends out a signal. The signal is not proportional to the temperature. It goes from off to on as the temperature passes the switching level.

Look at the photo on the left. The watches are both showing the same time. The analogue watch on the left shows the time by the distance the hands have moved round the dial.

These watches show the same time in different ways

The digital watch on the right shows the time as a set of numbers. A single digital signal is either on or off but several digital signals can be arranged together to form the number display you can see.

IT AND MICROELECTRONICS

20 years ago, most of the equipment in research laboratories was analogue. It included mercury thermometers, spring balances and electric current meters with dials. Today, much of the electronic equipment in laboratories is digital. This has one big advantage over analogue equipment. The measurements (signals) on analogue sensors had to be read, and the results were written down by hand. On the other hand, digital sensors can be connected directly to electronic systems and computers. These computers and electronic control systems use digital signals. This means that experiments can be controlled and monitored automatically.

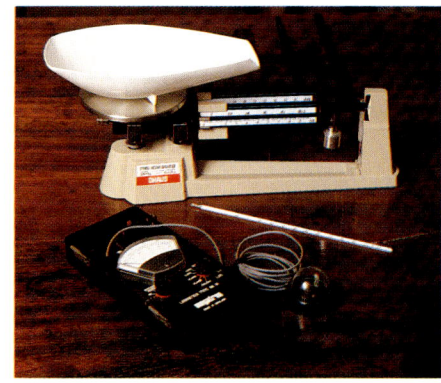

Some analogue equipment that was common in research laboratories

Things to do

1 Morse code is a simple digital code for sending messages by radio. Each letter is coded as a pattern of dots and dashes. A dot is a short pulse, a dash a long pulse. Here are the Morse codes for the letters of the alphabet:

a .—	j .———	s ...
b —...	k —.—	t —
c —.—.	l .—..	u ..—
d —..	m ——	v ...—
e .	n —.	w .——
f ..—.	o ———	x —..—
g ——.	p .——.	y —.——
h	q ——.—	z ——..
i ..	r .—.	

a) Why is Morse code a digital code?
b) With a friend, try out Morse code as a way of sending messages. Write a message in Morse code, and then send it to your friend. Your friend should 'read' the message and then reply with his or her own message, sent to you in the same way. Time how long it takes each of you to 'read' and 'write' the messages.
c) What are the advantages and disadvantages of using Morse code as a way of communicating?
d) Find out about the origins of Morse code. Who invented it, and when, and how was it used? Is Morse code still in use today? Write a short article about Morse code for your local paper. You might need to ask relatives, or use the library in your research.

2 Look at these photos. List the objects in the photos which produce digital signals and those that produce analogue signals.

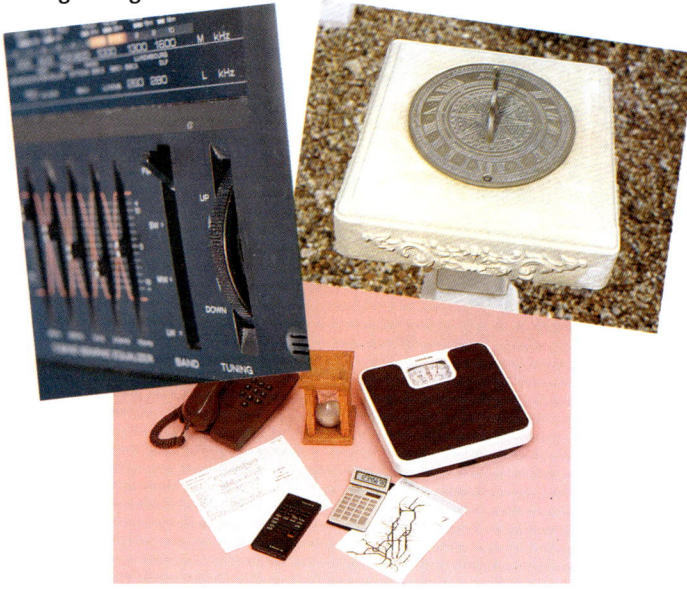

3 Which of the following could you measure using an *analogue* signal?
  a) Timing the winner of a race;
  b) counting items in a supermarket trolley;
  c) measuring the height of sunflowers in a sunflower-growing competition;
  d) timing the boiling of an egg;
  e) comparing the strength of different people.

4 Why do you think it is better to use digital sensors for laboratory experiments? What is one *disadvantage* of using digital sensors?

9 Logic gates and circuits

An electronics kit showing AND, OR and NOT gates

In the last unit, you saw that digital signals are either on or off. They have only two states. These states are also described as:

high or low,
1 or 0,
true or false.

The high (or 1, or true) state corresponds to a high voltage, and sometimes a current flowing in the circuit. The low state corresponds to a low voltage and no current.

When you are constructing control systems, it is useful to combine digital signals. You can do this by passing the signals through circuit components called **gates**.

A **gate** has one or two input signals, and produces a single output.

The output signal depends on the values of the inputs. You can use an electronics kit, like the one shown in the photograph, to construct circuits of this sort.

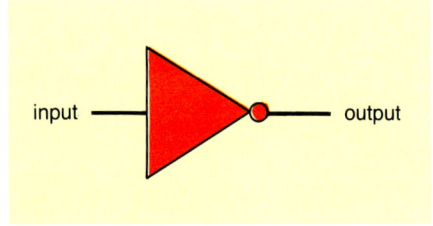

Figure 1 A NOT gate

NOT Gate

A gate called a **NOT gate** is shown in figure 1.

The simplest type of gate is a NOT gate. A NOT gate reverses the digital signal coming into it. If the input is 0, the output is 1. If the input is 1, the output is 0. This is shown in the table below for a NOT gate on the left. Tables like this are called **truth tables**.

Truth table	
Input	Output
0	1
1	0

AND Gate

One way of combining two digital signals is through an **AND gate** (figure 2). The output signal from an AND gate is 1 if one input *AND* the other input are both 1. All other combinations of the inputs give an output of 0. The truth table below shows all these possible combinations.

Truth table		
Inputs		Output
0	0	0
0	1	0
1	0	0
1	1	1

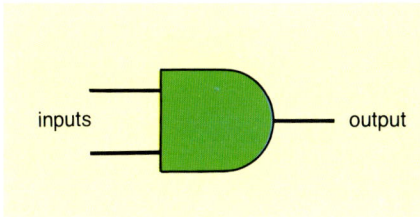

Figure 2 An AND gate

OR Gate

Another way of combining two digital signals is through an **OR gate** (figure 3). The output signal from an OR gate is 1 if one input *OR* the other is 1. If both inputs are 0, then the output is zero. The truth table below shows the possible combinations.

Truth table		
Inputs		Output
0	0	0
0	1	1
1	0	1
1	1	1

AND, OR and NOT gates are called **logic gates**. This is because there are logical rules for working out the output from the inputs. These can be written out as truth tables like the ones above.

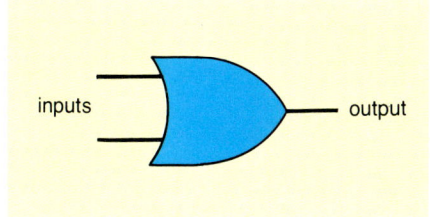

Figure 3 An OR gate

Two everyday devices which use logic circuits

LOGIC GATES AND CIRCUITS

Combining gates

AND, OR and NOT gates can be used together in circuits to control systems. These circuits are called **logic circuits**. For example, a control switch is 'on' (1) if one control input is 0 and the other is 1. The circuit for this is shown in the photograph and in figure 4.

An electronics kit set up as Output=NOT (A) AND B

The truth table for the logic circuit in figure 4 is made by combining a NOT gate with an AND gate. Check this truth table by making up the circuit with an electronics kit and trying out each combination of inputs in turn. Use switches to control the inputs, and connect the output to a light to see whether it is 0 (off) or 1 (on).

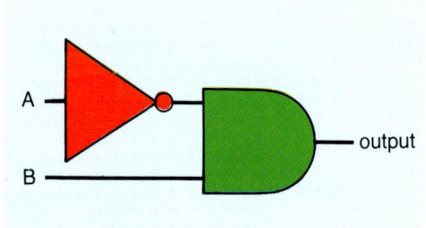

Figure 4 A logic circuit with a NOT gate and an AND gate

Truth table			
Input A	NOT (A)	Input B	Output
0	1	0	0
0	1	1	1
1	0	0	0
1	0	1	0

Circuits made up of AND, OR and NOT gates are used to control experiments and for many other purposes. You can use them in several of the activities later in this book.

IT AND MICROELECTRONICS

Things to do

1 a) Copy the truth table below and fill in the gaps.

Input A	Input B	A AND B	Output = NOT (A AND B)
0	0	0	1
0	1	–	–
1	–	–	–
–	–	–	–

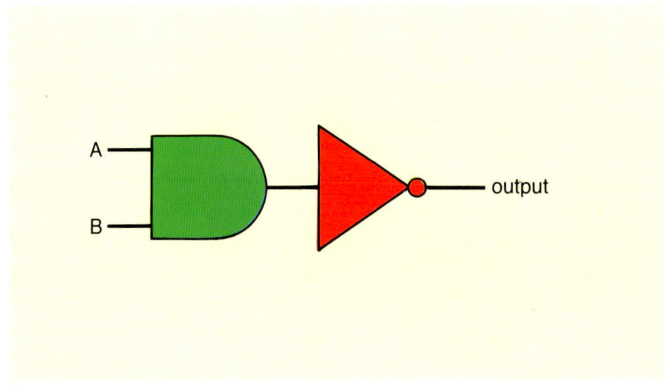

B Output=NOT(A AND B)

b) Draw a diagram of the circuit you would need to give the truth table in (a).
c) Connect up the circuit using your electronics kit to check the results in (a).

2 a) If one input to an AND gate is 1, how does the output relate to the other input?
b) If one input to an OR gate is 0, how does the output relate to the other input?

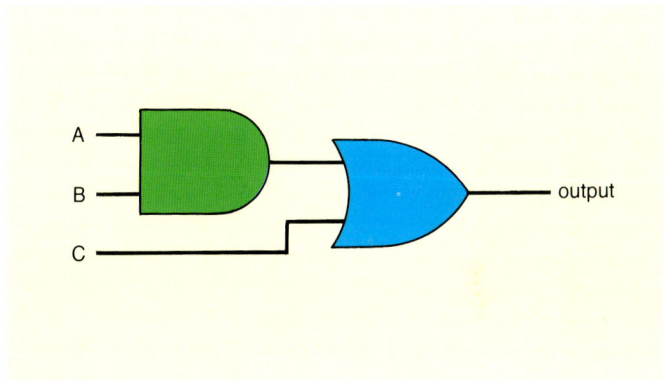

C Output=(A AND B) OR C

3 Using your electronics kit, make up each of the circuits shown in diagrams **A** to **D**. Draw up a truth table for each circuit, and check it against the results from the circuit you have made.

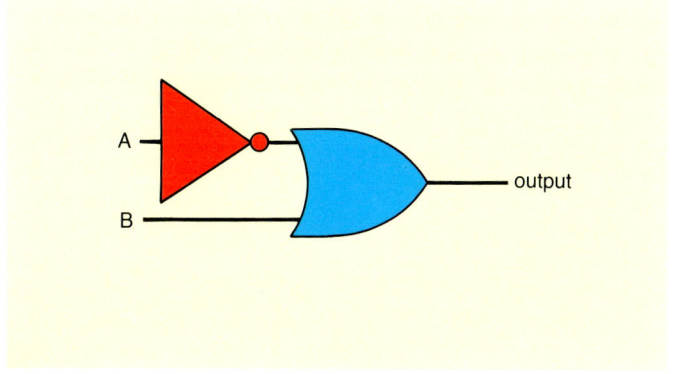

A Output=NOT(A) OR B

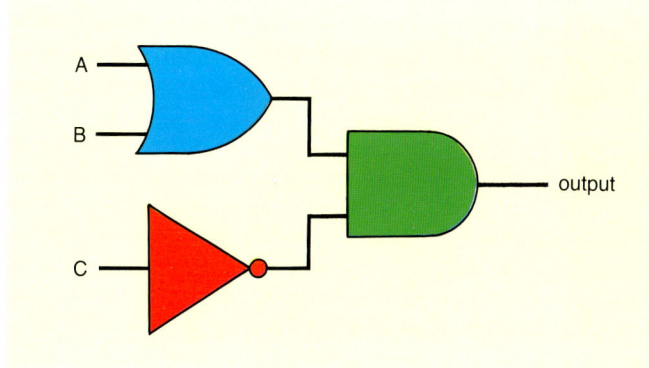

D Output=(A OR B) AND NOT (C)

10 Storing information

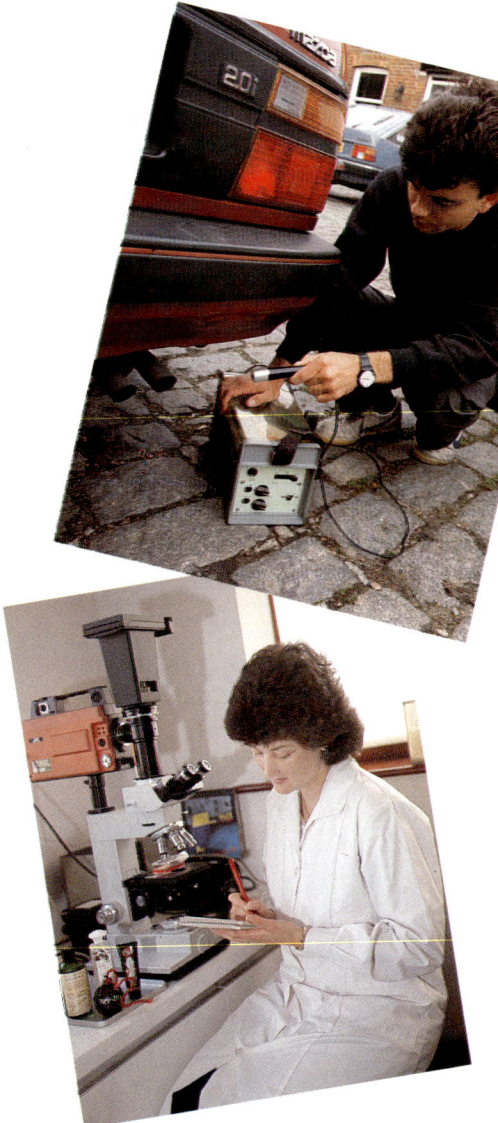

These three photos show different ways of storing scientific information. What are they?

When you do experiments or field work, you need to keep records. Some experiments produce a lot of results and information. These must be recorded carefully, so that they are not lost.

In the past, scientists used their notebooks to record their observations and results. They did all their calculations and drew their graphs by hand. This was a slow process and sometimes they made mistakes.

Today, we use many different devices to store scientific information. These include tape recorders, data loggers (which gather data from sensors), computers and even photos and videos. Using computers has many advantages:

- computers can store a lot of information in a very small space;
- they can do calculations quickly and directly on the stored information;
- they can display the information as tables or graphs;
- they can print information when it is needed;
- they can receive information directly from electronic sensors. This saves us having to type the information in, which is slow and which might introduce errors.

One way of storing information on a computer, and doing calculations on the information, is to use a **spreadsheet.** The photo on the left shows an example of a spreadsheet. A spreadsheet is a table, set out in rows and columns. Each entry in the table is called a **cell**. A cell contains a number or a word. You can enter a formula to calculate the number in one cell from other numbers in the table. You can save a spreadsheet on the computer's disk at any time, and print it. Most spreadsheets can be used to produce graphs from the numbers.

Spreadsheets have many uses. They are used in business and industry, as well as in science and engineering. Some spreadsheets are very large and complicated. Simple spreadsheets can easily be produced on your school computer. The main disadvantage is that it is not usually possible to get information directly from sensors into a spreadsheet. You have to type the information yourself.

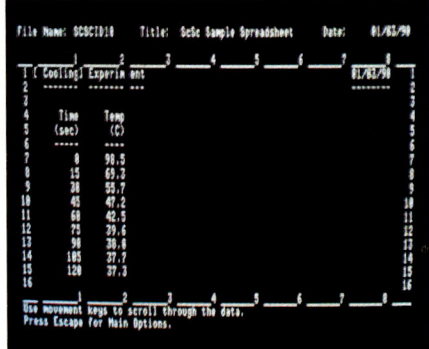

A spreadsheet showing the results of a simple experiment

24

Things to do

1 Working in a group with 2 or 3 others, discuss the most useful ways of storing the following types of information:
   a) interviews with people in the street;
   b) a record of the water level in a reservoir over a week;
   c) information about the way in which a car windscreen breaks when something hits it;
   d) the temperature and pressure of a chemical reaction at 5 minute intervals;
   e) information about the way an animal looks after its young.

2 Shereen and Joe wanted to investigate whether someone's mass is related to their height. One way of investigating this is to measure the height (in millimetres) and mass (in kilograms) of a group of students.

The measurements of height and mass for the 9 students shown in the drawing were taken. The measurements are in the table on the right. Open up a new spreadsheet on your school computer, and enter the results. (If you cannot use a computer draw up your own spreadsheet.)

Name	Height mm	Mass kg
Jason	1475	34.2
Darryl	1342	29.7
Ling	1376	31.2
Meera	1287	29.1
Cal	1341	30.8
Sara	1486	33.8
Paul	1134	26.7
Jon	1250	28.2
Mari	1330	29.5
Average		

At the bottom of the table, enter a formula in the cell for the average height, and one for the average mass. If you cannot do this, ask your teacher to help you work them out and type them in.

   a) Look at the results. Which pupils are:
      (i) above average height,
      (ii) above average mass,
      (iii) above average height and above average mass,
      (iv) below average height,
      (v) below average mass,
      (vi) above average height but below average mass,
      (vii) below average height but above average mass,
      (viii) below average height and below average mass?
   b) From the spreadsheet, plot the following graphs. (If you cannot plot graphs directly from the spreadsheet, draw the graphs by hand.)
      (i) A bar graph of the height and mass (*y*-axis) against the names (*x*-axis). Draw the bars in order of increasing height.
      (ii) A line graph of height (*y*-axis) against mass (*x*-axis).
   c) Look at your answers to part (a), and the graphs. Is there any pattern in them? Discuss whether there is a relationship between height and mass in this group of students.

Save your spreadsheet on disk and print a copy of it.

11 Communicating information

Look at the photograph above. It shows equipment which is used for experiments on a satellite.

Some of the most useful experiments in recent years have been carried out in space. The results are recorded on data loggers and computers in the satellites. They are then sent back to earth by radio waves.

This shows the importance of getting information from one place to another. This is called **communicating information.**

Communicating information is moving it from one place to another.

The telephone system allows you to communicate with other people anywhere in the world. Calls are connected by exchanges. Modern telephone exchanges, like the one on the right, are electronic. The telephone system also handles messages between computers, and telex and facsimile messages

26

IT AND MICROELECTRONICS

Laboratories are often linked to computers, or other laboratories a long distance away, by telephone lines. One of these links is shown in the photograph below. It is called a **modem**. Using these telephone links, information can be sent at high speeds. This enables laboratories to **share information** even though they are far apart.

Getting information from one place to another is also important on a smaller scale. The photograph on the right shows an experiment in an industrial research laboratory. The results are monitored electronically and fed straight into a computer. The computer stores the information until it can be used.

Communication links between laboratory equipment and computers make it possible to send large amounts of information quickly and accurately. The information is also in a form ready to be stored or processed. So, a lot of time and money is saved and there are no errors from typing the information into the computers.

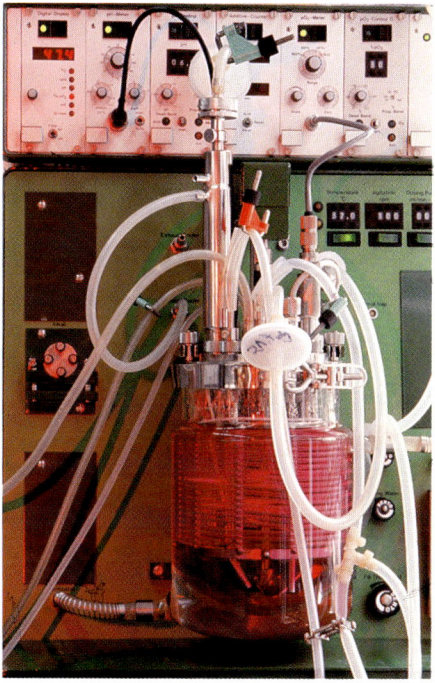

The information from this experiment is fed directly into the computer

A connection between a computer and a telephone line

Things to do

1 Sara works in the research laboratories of *Chemtech Industries*. Her laboratory is in Swindon. She has produced a set of results from an experiment. She needs to send them to Daniel, who works at the Birmingham office of the company.
   a) What sort of communication could Sara use? Explain your choice.
   b) What other ways can you think of to transfer information between the 2 offices?

2 Different types of information can be sent by communication links from one place to another. On page 84, you saw how information about changes in the weather is sent to weather forecasters to help them make their predictions. Make a list of other types of information which are sent by communication links. You can set out your list in a table like the one shown.

Information which is being sent	From	To	Communication link
air pressure for weather forecasts	weather balloon	weather station	radio

3 What are the advantages of:
   a) sending information directly from sensors to computers;
   b) long-distance communication between laboratories and computers?

27

12 Displaying information

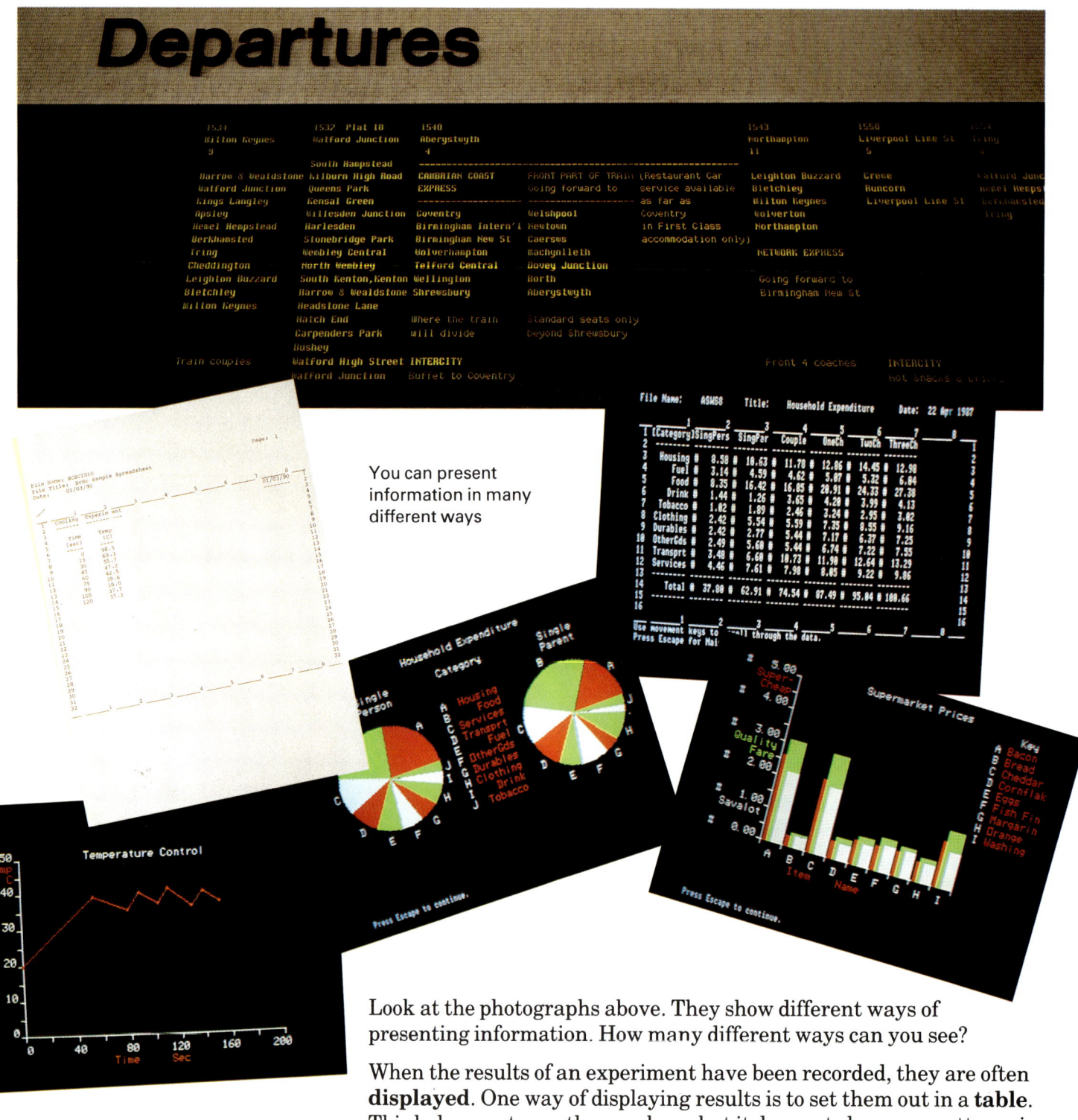

You can present information in many different ways

Look at the photographs above. They show different ways of presenting information. How many different ways can you see?

When the results of an experiment have been recorded, they are often **displayed**. One way of displaying results is to set them out in a **table**. This helps you to see the numbers, but it does not show any patterns in the results.

A much better way of showing patterns in results is to draw a graph or a chart. There are different types of graphs and charts. Line graphs, bar charts and pie charts are the commonest examples. The most suitable type of graph or chart depends on the nature of the results.

IT AND MICROELECTRONICS

Well-presented results:

■ make an experiment much easier to understand;

■ show any patterns in the results very clearly.

In research and industrial laboratories, graphs are usually produced by the computers which store and analyse the results. Some school computers can produce graphs directly from spreadsheets. With others you have to draw the graphs yourself.

Things to do

1 Dry air is made up of these gases:

Nitrogen: 78.09% Oxygen: 20.95%
Argon: 0.93% Carbon dioxide: 0.03%

a) What do you think is the best way to display this information?
b) If possible, use your school computer to produce your graph or chart. If this is not possible, you can draw it by hand.

2 During a chemical reaction, Suzi measured the temperature at 10-second intervals. Her results are as follows:

Time sec	Temp °C	Time sec	Temp °C
0	20	10	37
20	59	30	97
40	139	50	118
60	96	70	74
80	61	90	53

a) Redraw Suzi's table in a clearer layout.
b) What is the best way to show the pattern in her results?
c) Draw the graph or chart you have chosen.

3 Most people have hair that is either black, brown, red or blonde. Most people's eyes are either blue, green or brown. Taken together, this gives twelve groups:

   black hair + blue eyes,
   black hair + green eyes,
   black hair + brown eyes,
   brown hair + blue eyes, etc..

a) List all the groups.
b) Carry out a survey of all the people in your class. Count the number in each of these groups.
c) Decide on the best way to present your results: a table, a chart, a graph or some form of diagram. Explain briefly the reasons for your choice.
d) Present your results in the way you have chosen.
e) Compare your results with those of other people in your class.
f) Can you draw any conclusions from your results?

4 How would you display the following types of information? Think about how the information will be used, and who will use it.

a) The temperature of a patient in intensive care in hospital, measured every hour.
b) The results of a survey into the TV programmes watched on Thursday evening by the people in your class.
c) The final heights of 5 bean plants grown in different conditions for 2 weeks.
d) The average weekly rainfall in your school grounds over a term.
e) The speed of a bicycle, measured every minute, on a 30 minute journey.
f) The height of water in a reservoir, measured daily from June to September.

Activities

1 Rolling race

Do all round objects roll down a slope at the same speed? The aim of this activity is to find out.

Work in groups of 3 to 5 pupils.

You will need:

- 3 sticks, each 1 metre long,
- about 10 round objects such as a tennis ball, football, ball bearing, marble, roll of tape,
- digital stopwatch or similar timer,
- calculator,
- graph paper.

What you do

1 Tape the 3 sticks together to make the slope.

2 Support one end on a pile of books, for example, to give the metre sticks a slope of about 30 degrees. (The precise slope does not matter, as long as it is the same for all the objects.)

For each object:

- hold the object steady at the top of the slope;
- set the timer to zero, and have it ready to start;
- release the object and start timing;
- stop the timer as the object reaches the end of the slope;
- write down the time in your notebook.

Making a table of results

4 Draw a table for your results like this.

Rolling race	Distance: 1 metre	Date:
Object	**Time** sec	**Average speed** m/s
–	–	–

5 You can work out the **average speed** of each object by using this formula:

$$\text{average speed} = \frac{\text{distance}}{\text{time}}$$

Notice that you are calculating the *average* speed. This is because the objects started at rest and speeded up (accelerated) as they rolled down.

The distance is 1 metre for all the objects. Using a calculator, find out the speed of each object from the formula above. For example, if an object took 4.3 seconds to roll down the slope:

$$\text{average speed} = \frac{1.0\,\text{m}}{4.3\,\text{s}} = 0.23\,\text{m/s}$$

Work out all the speeds in this way, and copy them into the table. Round them to 2 significant figures.

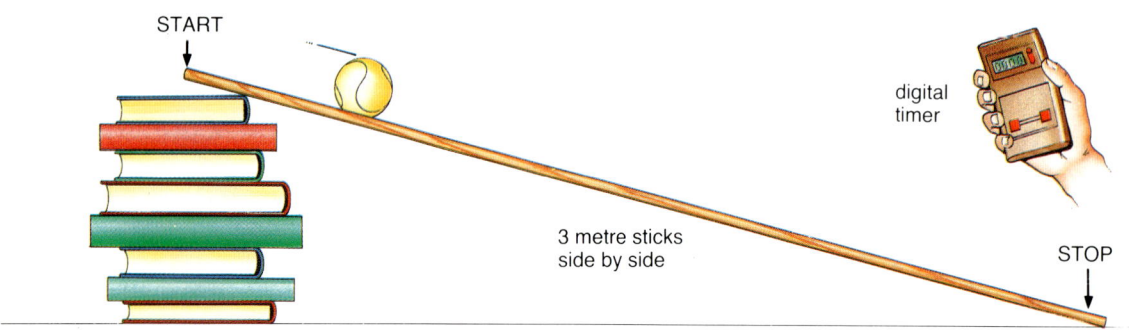

The rolling race in action

6 Use a bar chart to compare the average speed of each object. (The speed should be on the *y*-axis, and each object should have a bar along the *x*-axis.) It is a good idea to arrange the bars in order, fastest to slowest.

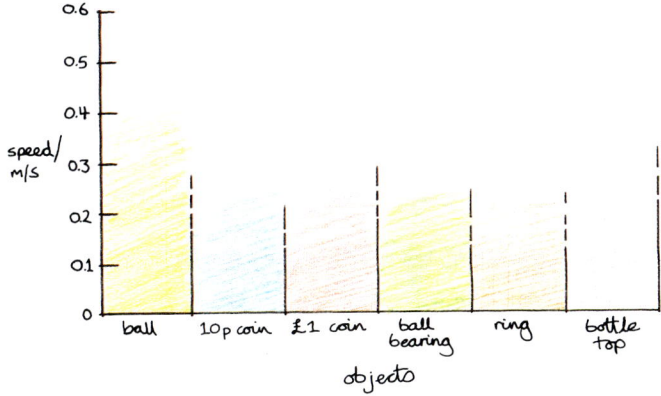

Questions

Look at your table of results and your bar chart.
a) Do heavy objects roll faster than light ones?
b) Do large objects roll faster than small ones?
c) Do objects with holes in the middle roll faster than solid ones?
d) Was your investigation a fair test? How could you have improved it?
e) Compare your results with those of other groups in your class. What reasons might there be for any differences in your results?

Your bar chart should be set out like this

2 Choosing an athletics team

Madhur Uddin is the coach of an athletics team, which is training for a major competition. The competition is now one week away, and she has to select a team for the 4 × 4 100 m relay.

All 10 members of the team run in three trials, each 100 m long. Madhur collected their times for each trial and wrote them down in her notebook. You can see her results on the right.

1 Copy out the table, and work out the average speed for each athlete.
2 Draw a bar chart of the average speeds.
3 Who would you select for the relay race?
4 Suggest an order in which the athletes should run, and explain why you have chosen this order.
5 Do you think it is right to select the athletes in this way? What other factors might be important? How could you improve this selection process?
6 What equipment has been used to:
  a) obtain the information?
  b) perform the calculations?
  c) display the results?
Say which of these items of equipment are electronic.

Name	Time (seconds) Trial 1	Trial 2	Trial 3
Judy	13.1	12.9	12.8
Shareen	14.1	13.5	13.7
Michelle	11.9	12.3	12.3
Lea	12.7	13.0	12.8
Katie	12.2	12.7	12.7
Jan	13.8	13.6	14.0
Petra	12.9	13.4	13.3
Clare	13.6	13.7	13.9
Meena	12.6	12.2	12.2
Deborah	13.6	13.5	13.9

ACTIVITIES

3 Radar and echo sounders

Radar works by bouncing waves off objects

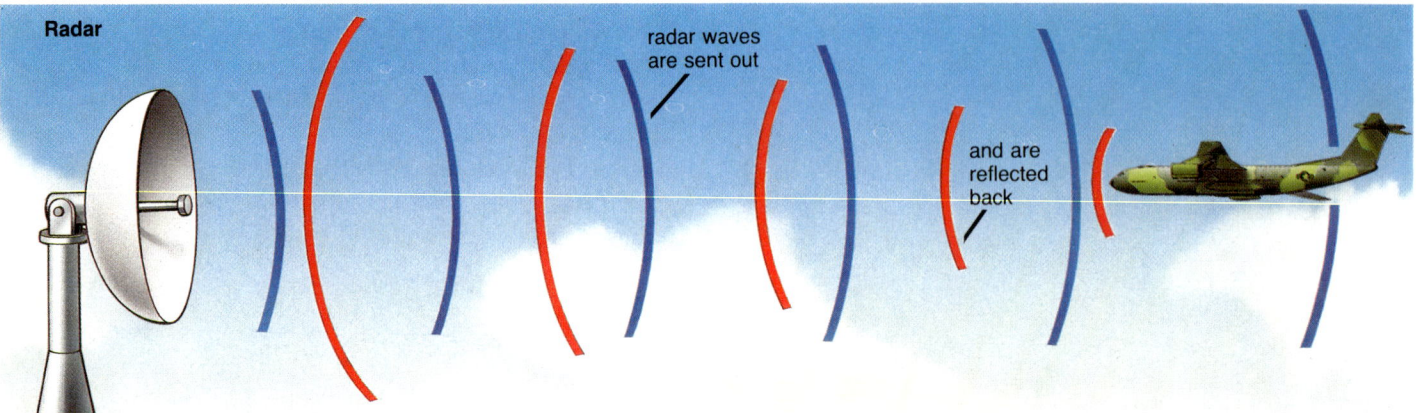

Various electronic systems can help us to measure speeds, distances and times. **Radar** is one of these systems. It is used by aircraft, ships and submarines every day. It gives them information about what is around them so that they can navigate, even if the crew can't see what is ahead.

Radar works by sending out pulses of **radio waves**. When they hit a solid object, they bounce back. The time taken for the radio waves to travel from the radar to the object and back again tells you how far away the object is. The radar system is complicated but the principle on which it works is simple. Using electronics, radar can calculate distances automatically and display the results on a screen like the one below.

A radar screen can show us how far away things are

Echo sounders work in a similar way to radar but they use **sound waves**. You can find out the depth of the sea using the time taken for these waves to travel through the water. An echo sounder can also find submarines, and even fish.

1 Describe (using diagrams) how an echo sounder in a boat would detect a shoal of fish.
2 Estate agents have to measure the size of each room in a house or flat that is for sale. They used to use tape measures to do this, but today they have electronic devices. How do you think these devices might work? What advantages would an electronic device have over a tape measure?
3 Look at the photo below. Ultrasonic scanners are used to look at a baby inside the mother's womb. The scanner works like an echo sounder. It uses high pitched sound waves which people cannot hear. Discuss all the benefits of using ultrasonic scanners in this way.

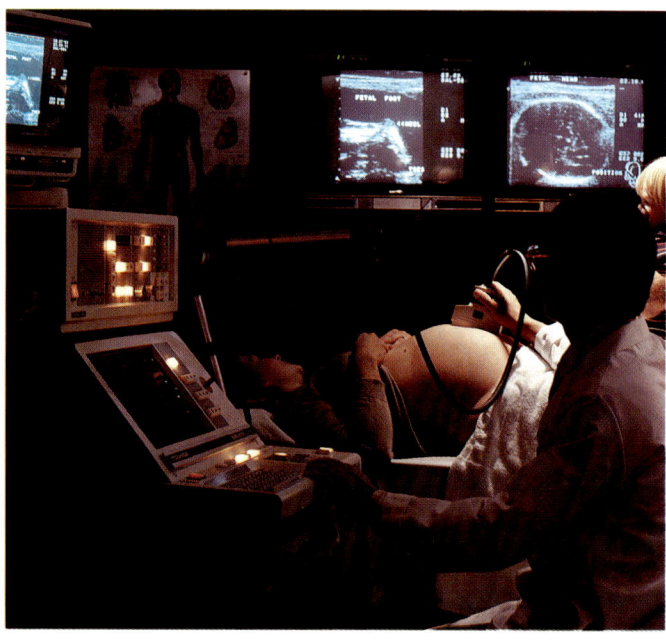

4 Monitoring a chemical reaction

This apparatus is used to produce a vaccine for hepatitis. The chemical reactions which produce the vaccine need to be monitored all the time

In laboratories, many reactions need careful monitoring. Monitoring provides information about a reaction. We use this information to ensure that the reaction takes place safely. We also use the information to control a reaction. In this activity, you will carry out a simple chemical reaction and monitor it.

Work in groups of 3 to 5 students.

You will need:

- a beaker (about 500 cm^3),
- water, sugar and dried yeast,
- a digital top pan balance,
- a digital temperature probe or thermometer,
- graph paper,
- a stopwatch or digital timer,
- a plastic stirrer (*not* metal).

What you do

1 Pour 250 cm^3 of warm water at about 45°C into your beaker.
2 Add 10 g of sugar and stir it until it has all dissolved.
3 Add 30 g of dried yeast and stir it until it has all dissolved. Allow a few minutes for the yeast to begin to react with the sugar. As this happens, a froth forms on the surface of the liquid.

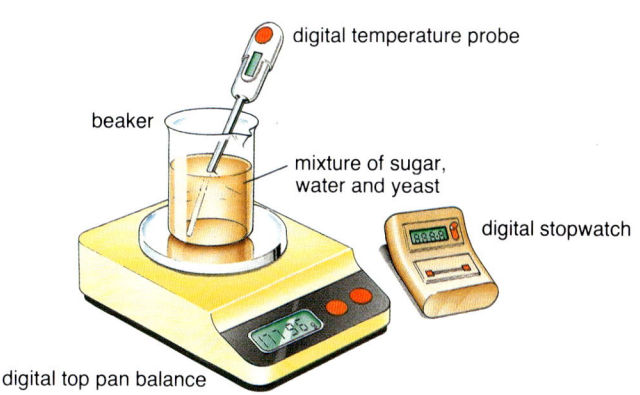

A beaker of reacting yeast on a top pan balance

4 Once the reaction has started, place your beaker of reacting yeast on the top pan balance, and place the temperature probe in it.
5 Start the timer and write down the temperature and the mass in your notebook.
6 Make a note of the mass and temperature every minute for the next 15 minutes. Remember to stir the liquid from time to time.
7 Set out your results in a table like the one below.

Time (minutes)	Mass (g)	Temperature (°C)
0	—	—
—	—	—

8 a) Draw a line graph of temperature (on the *y*-axis) against time (on the *x*-axis).
b) Draw a graph of mass (*y*-axis) against time (*x*-axis). Use the diagrams below to help you set out your graphs.

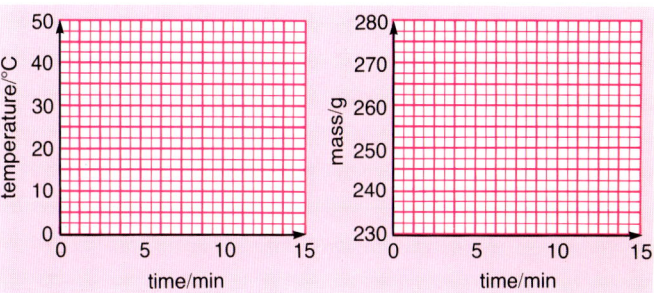

Your graph of temperature against time should be set out like this

Your graph of mass against time should be set out like this

ACTIVITIES

Questions

Look at the results in your table and your graphs.

a) What happens to the temperature during the reaction?
b) Is heat given out or taken in during the reaction?
c) What happens to the mass during the reaction?
d) Suggest a reason for the change in mass.

Taking it further

a) If you have a data logger, you could use it to monitor the temperature. Transfer the results to a computer to produce the graph of temperature against time. (You will still have to monitor the mass by writing the figures in a table).
b) Although you have used an electronic balance, timer and temperature probe (and possibly a data logger), you have still monitored this reaction by hand. What would it be like if the reaction went on for several hours (or even days)? What are the advantages and disadvantages of monitoring chemical reactions automatically?
c) Imagine carrying out this investigation using analogue equipment. What difference would it make? What are the advantages of using electronic equipment in this investigation?

5 Pulse rates

Your pulse tells you how fast your heart is beating. Does everyone's heart beat at the same rate? Does your heart beat at the same rate all the time? In this activity you will investigate these questions.

The activity can be done by the whole class working together, or in groups of about 10.

You will need:

- a computer,
- a spreadsheet program,
- a program which can plot graphs from spreadsheets (if the spreadsheet program cannot do this),
- a watch or electronic timer.

What you do

1 Make sure that you can take your own pulse. The best way is at your wrist.
2 Get someone with a watch or timer to call out the start and end of a minute. Take your pulse during this time.

3 Start the spreadsheet running, and take it in turn to enter your name and pulse rate into it. Follow the example in the photograph below.

Your spreadsheet of pulse rates should look like this

4 If the spreadsheet has a sort facility, you can now sort the rows in order of pulse rate.
5 At the bottom of the spreadsheet, enter a formula for the average pulse rate.
6 Turn the calculation on, and check that the average appears correct.

Producing a histogram

The best way to see the spread of pulse rates is to produce a histogram. This is a bar chart of the number of people whose pulse rates fall within a series of intervals.

7 Look at the pulse rates, and find the smallest and largest rates.

8 Choose a set of about 5 equal intervals which includes the smallest and largest. For example, if your results range from 77 to 96 beats per minute, you could choose the following intervals:

    75 to 79
    80 to 84
    85 to 89
    90 to 94
    95 to 99

9 Enter the intervals in two columns of the spreadsheet, as shown in the photograph.
10 Now count the number of people in each interval, and enter these numbers (the frequencies) in the next column.
11 Plot a bar graph of the intervals on the *x*-axis against the frequencies on the *y*-axis.
12 Print the bar graph.

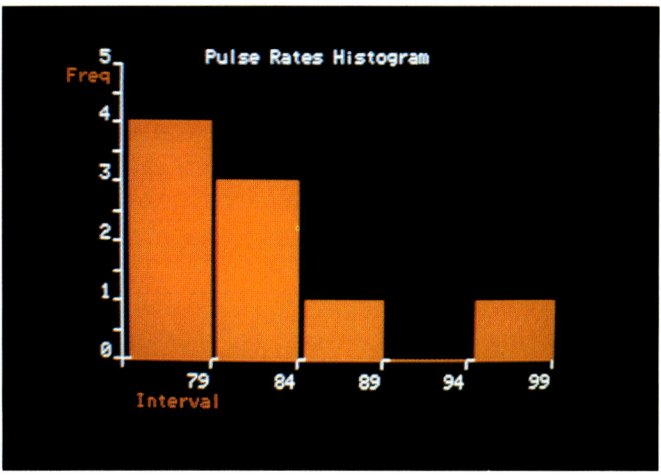

Your histogram should look something like this

Questions

Look at the shape of your histogram.

a) Is there one 'peak' in the histogram, or more than one?
b) What does the 'peak' in the histogram indicate?
c) Are most people's pulse rates near the average?
d) Suggest some reasons for the spread in pulse rates.
e) Repeat the pulse rate investigation after doing some exercise. What differences does this make to the results?

6 Electrocardiographs

When you take your pulse, as in the last activity, you are counting the number of beats per minute. An electronic device called an electrocardiograph (ECG) measures the *pattern* of a person's heartbeat, as well as the number of beats per minute. This gives detailed information about the way the heart is working. An ECG can even be used for a baby before it is born.

1 Find out more about the uses of an ECG. What are its benefits?
2 How many uses of pulse rate measurements can you think of?
3 Look at the 3 traces on the right. Compare the 2 abnormal traces with the normal trace and point out any differences and any similarities.

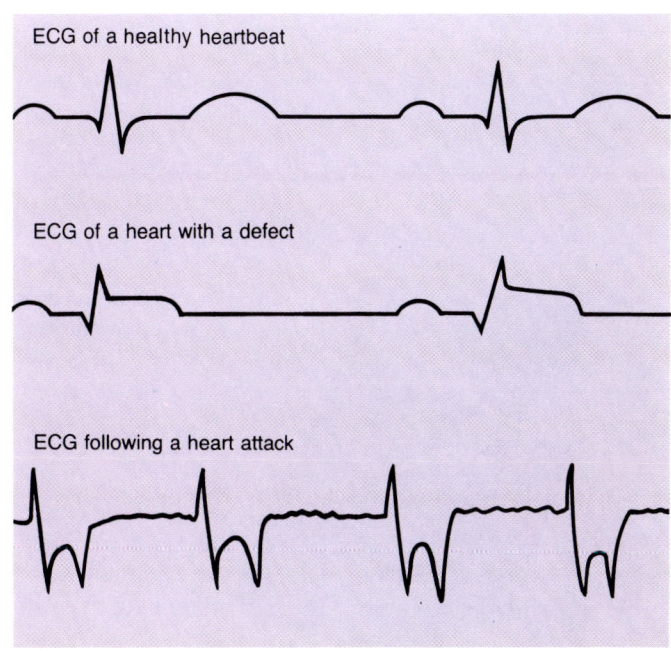

ACTIVITIES

7 Rapid reactions?

If someone steps in front of your bicycle, how soon do you put on the brakes? If the traffic lights change to red, how quickly does a driver slow down? If the pilot of an aircraft suddenly sees another aircraft approaching, how quickly does he turn his plane away? These examples show how important our **reaction** or **response** time is. It is often essential in an emergency. In this activity, you can investigate your own reaction (response) times using a simple electronic circuit.

Work in groups of 4 or 5 students.

You will need:

■ an electronics kit including a counter, switches, a NOT gate, an OR gate, a light and a buzzer,

■ a spreadsheet and graphics program running on a computer.

The steps

1 Connect the battery to your electronics kit and check that it is working.

2 One of the circuits in the kit is a counter. It displays the binary and hexadecimal (base sixteen) numbers:

Binary	Hexadecimal
0000	0
0001	1
0010	2
–	–
1111	F

An electronics kit set up as a timer

Press the reset switch to make the counter go to 0000. Then click the count switch up and down to check that it counts through the numbers. When it reaches 1111, it should return to 0000 and keep going. If you connect the pulse unit to the count input, it should count through the numbers automatically.

3 Connect up the logic circuit shown in the diagram and described below. It converts the counter into a timer with one switch to start it and one to stop it.

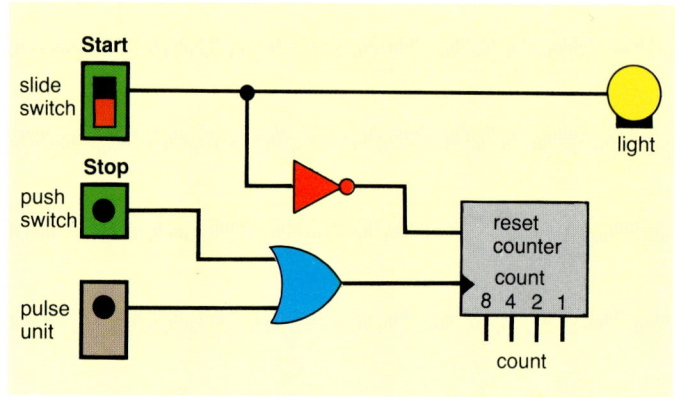

Logic circuit for a reaction timer

■ Connect the start switch to a light. Also connect it to the input of a NOT gate.

■ Connect the output of the NOT gate to the reset input of the counter.

■ Turn the pulse unit to its fastest speed, and connect it to an input of the OR gate.

■ Connect the stop switch to the other input of the OR gate.

■ Now connect the output of the OR gate to the count input.

4 Check the connections and test the circuit.

■ When the start switch is off, the count should be 0000 and the light should be off.

■ When the start switch is on, the light should be on, and the counter should start.

■ When the stop switch is pressed, the counter should stop. Its count should show where it stopped.

5 Take turns to see how long it takes you to press the stop switch after you see the light come on: one member of the group slides the start switch. The one whose response time is being tested presses the stop switch when he or she sees the light come on. The count shows how long he or she has taken to respond.

Let each member of the group have about 5 turns. Write down the names and times in your notebook.

Recording the results

6 Start the spreadsheet program on the computer, and open up a new spreadsheet for your results.

7 Take turns to enter the names and times you have written down, as shown in the photograph.

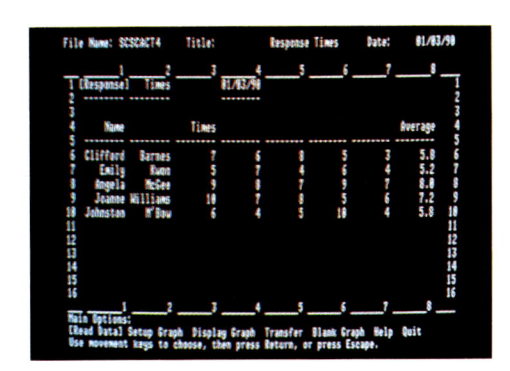

Set out your spreadsheet like this

8 Enter formulas in the last column to work out the average time for each pupil. Ask your teacher to check this.

The best way of showing the spread of the times is to use a histogram (a bar chart).

9 Look for the shortest and the longest reaction times, and then enter the whole range of times in a new column of the spreadsheet. For example, if the shortest time is 3, and the longest is 11, enter the numbers 3, 4, 5, etc., up to 11 in the column.

10 Now count the frequency of each reaction time, and enter the frequencies next to the times. For example, if there are five 3's in the table, then the frequency of 3 is 5. (Do not include the averages in counting the frequencies).

11 Produce a histogram of the times. This is a bar chart with the times as the x-axis, and the frequencies as the y-axis.

12 Print the spreadsheet and the histogram.

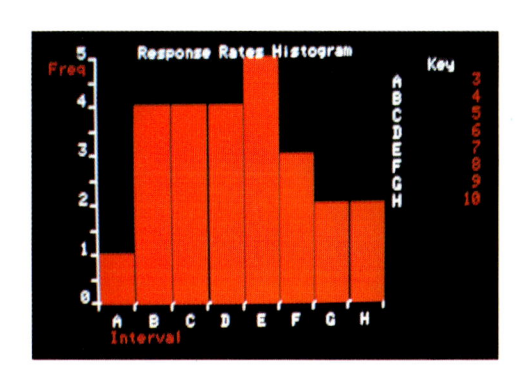

An example of a histogram of the response times

Questions

Look at the table of results, and the histogram, and answer the questions below. Remember that the times are not measured in seconds.

a) What is the value of the slowest time and the fastest time?

b) Is there one response time which is more frequent than the rest? If so, which is it? If there is more than one most frequent response time, are they next to each other?

c) What is the shape of your histogram? Why do you think it is this shape?

d) You need rapid response times in order to play some computer games successfully. Compare the response times of people in your class who regularly play computer games with those who don't. Are there any differences? Can you explain any patterns that you see? Are there any other games or sports in which response times are important?

Taking it further

a) Instead of connecting the start switch to a light, connect it to a buzzer. Close your eyes while you are testing your response to the buzzer. Repeat the experiment and compare the results with the buzzer to those with the light.

b) Connect the start switch to lights of different colours, or to circuits which produce other sounds. Do different colours or different sounds affect your response times?

c) Aircraft pilots must have good reaction times in order to ensure the safety of their passengers, crew and aircraft. Find out about the tests that pilots have to take before they qualify. How often do they have to repeat these tests after they have qualified? Prepare a 3 minute talk on this subject for the rest of the class.

ACTIVITIES

8 Counting made easy

An automatic traffic counter at the roadside

What you do

1 Connect the battery to your electronics kit and make sure that it is working properly before you start the activity.

This electronics kit has been set up as an event counter

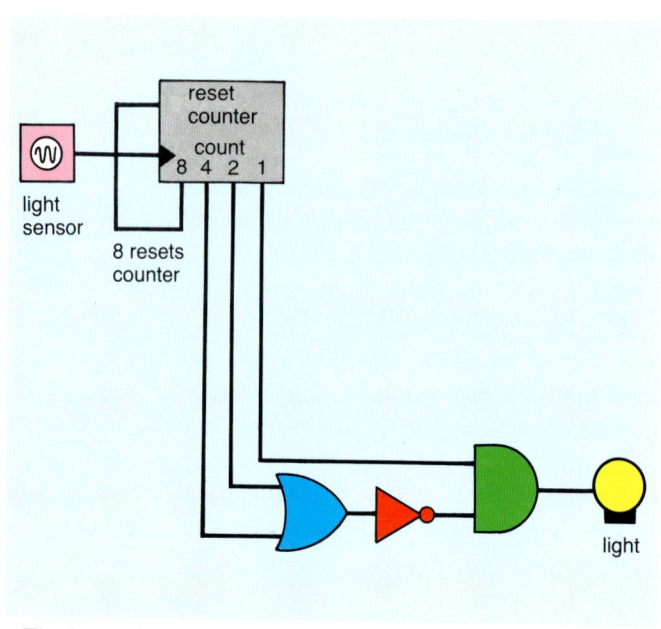

We often need to count the number of times something happens, such as the number of vehicles passing a junction, or the number of products on a factory production line. Counting events like this manually can be tedious and inaccurate. It is quite easy to set up an electronic counter to do this automatically. In this activity you will set up a simple event counter.

Your event counter will count repeatedly from 0 to 7. It switches on a light every time it is at 1. The counter is activated by a light sensor or a pressure switch.

Work in groups of 3 to 5 students.

You will need:

- an electronics kit with remote light sensor and pressure switch counter, AND, OR and NOT gates, and a light,
- cardboard, scissors and tape to make a small tube.

The logic circuit for an event counter

38

2 Connect the light sensor to the remote sensor socket, and connect this to the count input.
3 Connect the 8 output from the counter to the reset input. This restricts the counter to the range 0 to 7.
4 Connect the 2 and 4 counter outputs to the OR gate, and its output to the NOT gate.
5 Connect the output from the NOT gate and the 1 output from the counter to the AND gate.
6 Connect the output from the AND gate to the light.
7 Copy and complete a truth table like the one below for the circuit.

Inputs					Output
4	2	1	4 OR 2	NOT (4 OR 2)	NOT (4 OR 2) AND 1
0	0	0	0	1	0
0	0	1	–	–	–
0	1	0	–	–	–
0	1	1	–	–	–
1	0	0	–	–	–
1	0	1	–	–	–
1	1	0	–	–	–
1	1	1	–	–	–

8 By covering and uncovering the light sensor, make the counter cycle through the inputs in the table. Check that your table matches the output.
9 Make a small cardboard tube and place the light sensor in one end, so that it only detects light coming from one direction.
10 Now aim the tube across the desk, and adjust the dial so that the light sensor detects an object passing close to it.
11 Move an object, such as a small book, past the light sensor a few times. Check that the counter increases each time the object passes, and that the light goes on after every eighth pass.
12 Now replace the light sensor by the pressure switch. Support the pressure switch between 2 books, and roll a round object over it a number of times. Again check that the counter increases each time an object passes, and that the light goes on after every eighth pass.

Taking it further

The event counter you have made is limited by the logic gates available in your electronics kit, and what can be done in the classroom. Here are a number of additional suggestions you could try.

a) By using longer cables on the light sensor or pressure switch, set the counter up to count the number of students passing along a corridor, or the number of birds visiting a bird table, or the number of insects caught in an insect trap.

Your counter can be adapted to count the number of birds visiting a feeding log

b) If you have 3 AND gates, you can set up the counter to cycle through its whole range (0 to 15). You should make it switch the light on every time it reaches 15. Work out the logic circuit you would need.
c) Electronic counters like the one you have made are used for a lot of things. Look at the 3 examples below and then answer the questions.

- The numbers of cars entering and leaving a multi-storey car park are counted electronically. When the car park is full, no more cars are allowed to enter.
- Counters at the entrances to road bridges and tunnels can give us information on the amount of traffic that uses a particular route.
- Some London Underground stations have counters on automatic ticket gates. These count the number of people who pass through the gates.

(i) Why is it good to use automatic counters in each of these examples?
(ii) For each example, think of another method of counting which you could use.
(iii) How many other uses of automatic counters can you think of?

ACTIVITIES

9 Keeping warm

The clothes this skier is wearing must keep in as much heat as possible in the cold air

Have you noticed that some clothes keep you warmer than others? This is partly because some fabrics are better **insulators** than others (heat does not flow through an insulator very well). In designing clothes for different uses, it is important that designers know about the properties of the fabrics available. To do this, they might need to test the flow of heat through fabrics.

In this activity, you will investigate how heat flows through different fabrics. You could also look at other materials, such as plastics and metals.

Work in groups of 3 or 4.

You will need:

- a large beaker,
- hot (not boiling) water,
- a small test tube,
- a temperature probe or thermometer,
- a digital timer or stopwatch,
- a spreadsheet and graph display program on a computer,
- a variety of materials to test, such as aluminium foil, cloth, flat plastic sheeting, plastic bubble sheet packaging; (the test is done in water, so do not include paper or anything that will dissolve),
- a supply of small rubber bands.

What you do

For each material you are testing, the steps are the same.

1 Cover the test tube with the material being tested. Put a rubber band around it to hold the material in place.
2 Fill the test tube with hot water and place the temperature probe in it. Make sure that the water starts at the same temperature for each test. Remember to write down this temperature in your notebook.
3 Fill the beaker with cold water.
4 Place the covered test tube inside the beaker of cold water, and start the timer.
5 Note down the time and temperature inside the test tube every minute for the next 10 minutes.

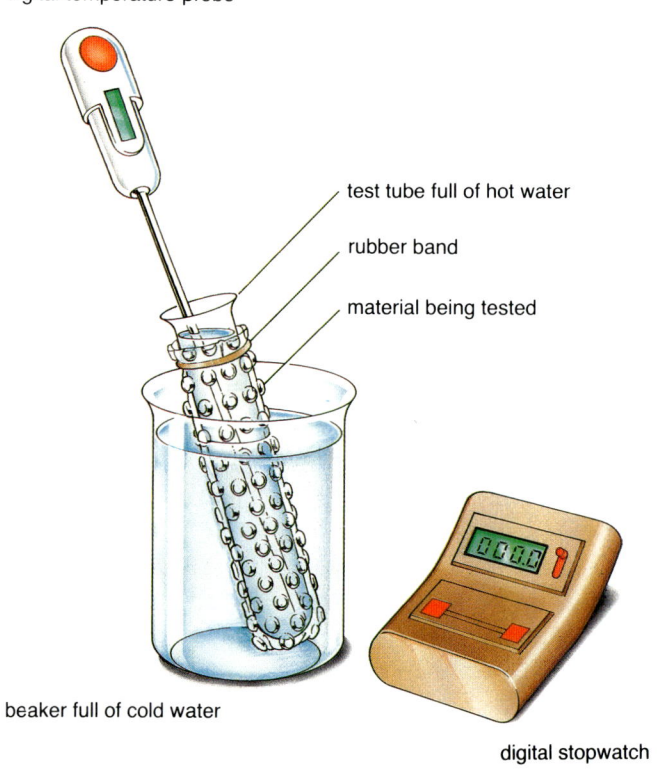

Testing the heat flow through plastic bubble sheeting

Recording the results

6 Start the spreadsheet program running and open up a new spreadsheet.
7 Enter the times and temperatures, using a column for each material you test.
8 Produce line graphs of temperatures (on the *y*-axis) and times (on the *x*-axis). If possible, put 3 sets of temperatures on the same line graph.

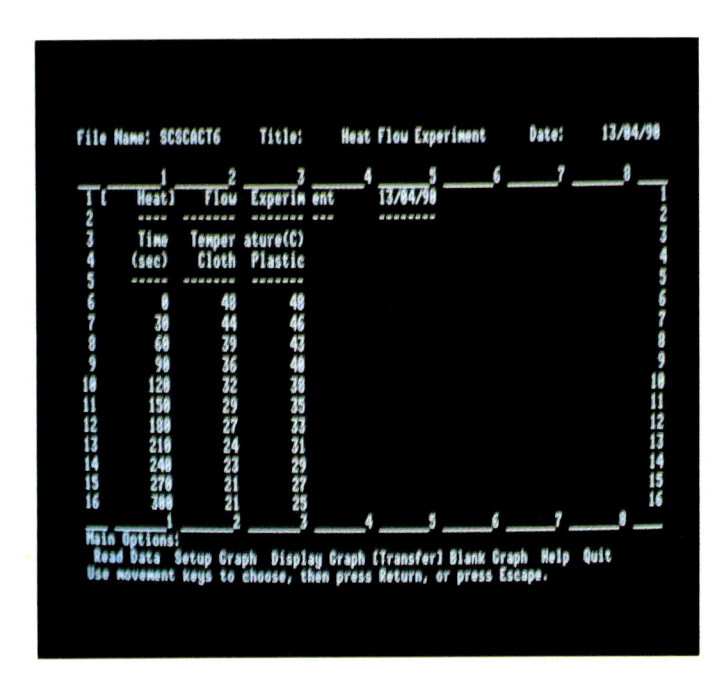

An example of a spreadsheet of results for cloth and plastic

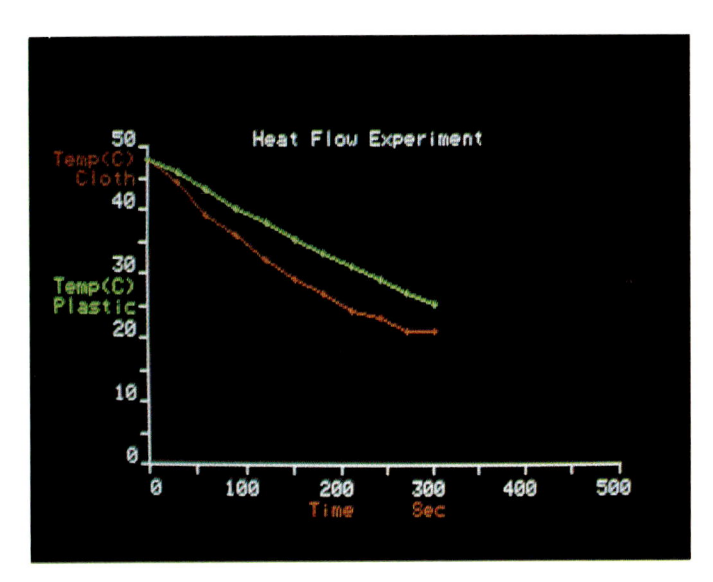

A line graph of the temperatures and times for cloth and plastic

Questions

Look at your results, and the graphs produced from them.

a) Which material took the longest to cool down?
b) Which material took the least time?
c) Which material is the best insulator?
d) Did the test tube cool down more quickly at the start of each test, or at the end? What reasons can you think of to explain this?

Taking it further

a) Repeat some of the tests, using 2 or 4 layers of the materials. What difference do more layers make?
b) This photo shows examples of objects in which the heat flow through the material is important.

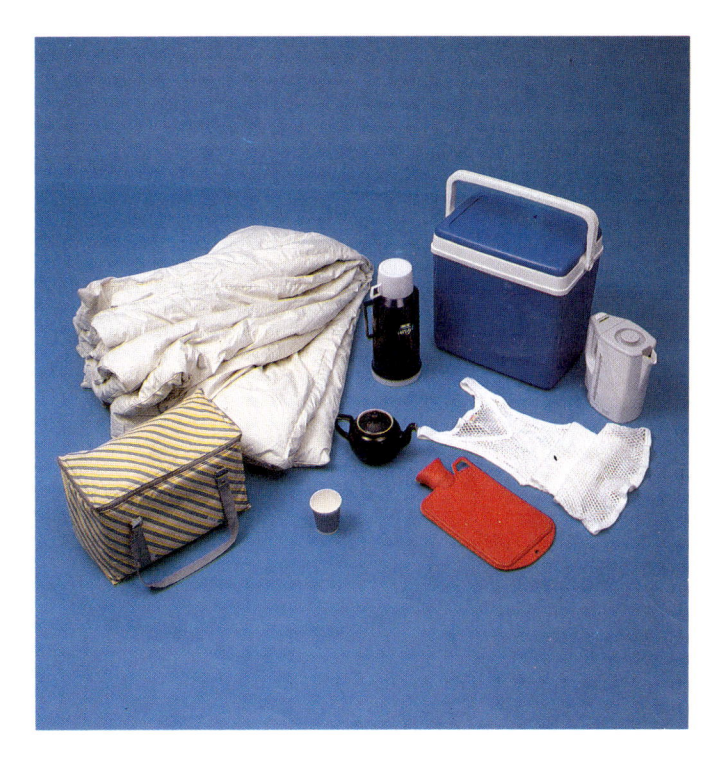

1 For each object, explain whether the material used is an insulator or not, and describe how this relates to the use of the object. Make a table of your comments.
2 Are there any other features of the design which help to control the heat flow?
3 Design an experiment to investigate how efficiently 2 or 3 of the objects in the photo work as insulators. You should try to use some electronic equipment in your investigation.

ACTIVITIES

10 Marks and spaces

Why is it hotter in summer than in winter? The main reason is that in summer, days are long and nights are short. In winter, it is the opposite: days are short and nights are long. We could draw a graph of this like the one below.

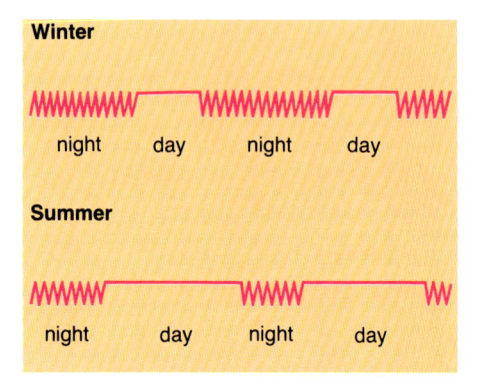

This is an example of **marks** and **spaces**. The ratio of the mark (the day) to the space (the night) is the **mark:space ratio**. The higher the mark:space ratio, the hotter the weather.

Control signals often use mark:space ratios, although they switch on and off more often than twice a day!

If the signal is on as long as it is off, the ratio is 1:1. If the signal is on for a quarter of the time, and off for three quarters, the ratio is 1:3.

In this activity you can create a signal with various mark:space ratios. You can then use it to control the heating of a flask of water. The temperature at the different mark:space ratios is measured and compared.

Work in groups of 3 or 4.

You will need:

- an electronics kit with a counter, a pulse unit, an AND, an OR and a NOT gate, a light and a relay switch,
- a battery-powered immersion heater which can safely be operated by the relay switch,
- a flask of water,
- a temperature probe or thermometer.

You should set up your heater circuit like this

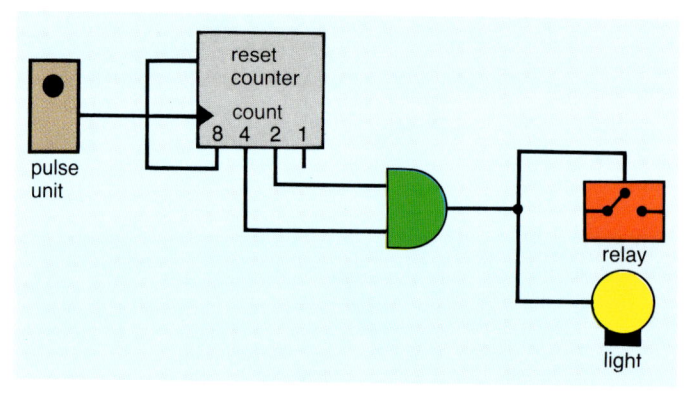

The circuit for an electronics kit connected to produce a signal with a mark : space ratio of 1:3

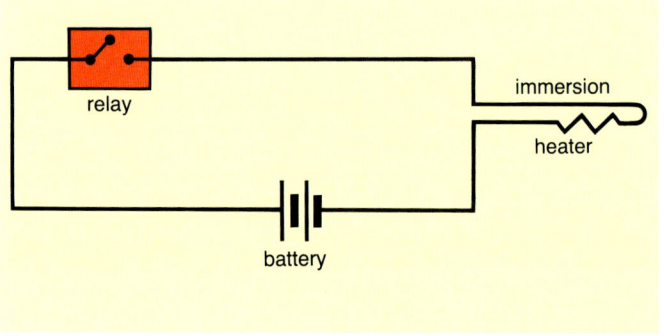

The circuit which heats the flask of water

42

IT AND MICROELECTRONICS

What you do

**Safety!
Do not get any water on the electric circuits!**

1 Connect the battery to the electronics kit and make sure that it is working.
2 Connect the pulse unit to the counter, and the 8 output from the counter to the reset input. Check that this makes it count from 0 to 7.
3 Then connect the 4 and 2 outputs from the counter to the inputs of the AND gate. Connect the output of the AND gate to the relay switch and to the light.
4 Next check that the circuit works like this:

Count	Light	Relay
0000	off	open
0001	off	open
0010	off	open
0011	off	open
0100	off	open
0101	off	open
0110	on	closed
0111	on	closed

The light is on for 2 counts and off for 6. This gives a mark:space ratio of 2:6 or 1:3.

Turn the pulse unit to a slow speed to check the circuit. Then turn it to its maximum speed.

5 Fill the flask with cold water. Measure the temperature of the water and write it down in your notebook.
6 Connect the heater circuit through the relay switch, as shown in the photograph and the diagram on the previous page, and place the immersion heater in the flask. When the relay is closed, the heater comes on.
7 Leave the heater to run for a few minutes at a mark:space ratio of 1:3. Measure the temperature from time to time, until it has reached a steady level. You should then note this level.

Your teacher will probably have to help you with the next few steps.

8 Disconnect the wires from the counter outputs, and work out how to get a mark:space ratio of 1:1. (Running the counter slowly and looking at the outputs will probably make it clear how to do it.) Reconnect the circuit for this ratio. Draw a sketch of the circuit you are using.
9 Leave the heater to run for a few minutes, at the mark:space ratio of 1:1. Make a note of the temperature the water reaches.
10 Once again, disconnect the wire(s) from the counter output(s), and work out how to get a mark:space ratio of 3:1. Reconnect the circuit for this ratio, and make a sketch of it.
11 Leave the heater to run for a few minutes at the mark:space ratio of 3:1. Make a note of the temperature the water reaches.

Displaying the results

12 You can display your results clearly using a table like the one below.

Mark:space Ratio	Temperature	Temperature Difference
No heating	–	0
1:3	–	–
1:1	–	–
3:1	–	–

The last column shows the difference between the temperature of the water at that point and the temperature of the water at the start, with no heating.
13 Plot a line graph of the temperature differences against the mark:space ratios.

Questions

Look at the information in your table and on your graph.

a) How does the temperature change as the mark:space ratio increases?
b) Is the change in temperature the same for each increase in mark:space ratio? Try to explain your results.
c) The heater is on 3 times as long with a mark:space ratio of 3:1 as it is for one of 1:3. Is the temperature difference 3 times as big? Try to explain your answer.
d) Modern electric trains are controlled by mark:space ratio signals. Find out about other examples of this kind of control, and discuss them.

ACTIVITIES

11 Keeping tropical fish

Tropical fish live naturally in warm water around the world. They cannot survive in cold water. If the temperature falls much below about 24°C, these fish would soon die.

This heater controls the temperature in the fish tank. The ideal temperature for most tropical fish is about 24°C

If you want to keep tropical fish at home, then you must control the temperature of the water in the tank carefully using a heater. The easiest way to do this is:

- turn the heater *on* if the temperature falls too *low*;
- turn the heater *off* if the temperature gets too *high*.

An electronics kit set up as a temperature controller

In this activity, you can set up a logic circuit which controls the temperature of a flask of water in this way. You could use this circuit to control the heater in a fish tank.

Work in groups of 3 to 5 students.

You will need:

- an electronics kit with a remote temperature sensor, a NOT gate, an AND gate, a relay switch and a light,
- a battery-powered immersion heater which can safely be controlled by the relay switch,
- a flask of water,
- a thermometer or digital temperature probe,
- a stopwatch or digital timer,
- a spreadsheet and graph display program on a computer.

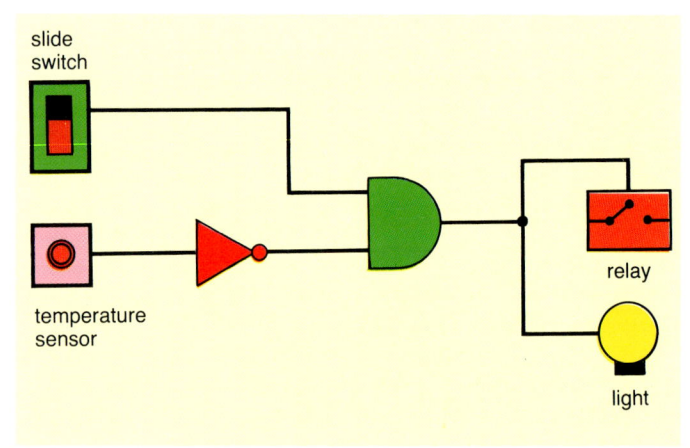

The logic circuit for controlling temperature

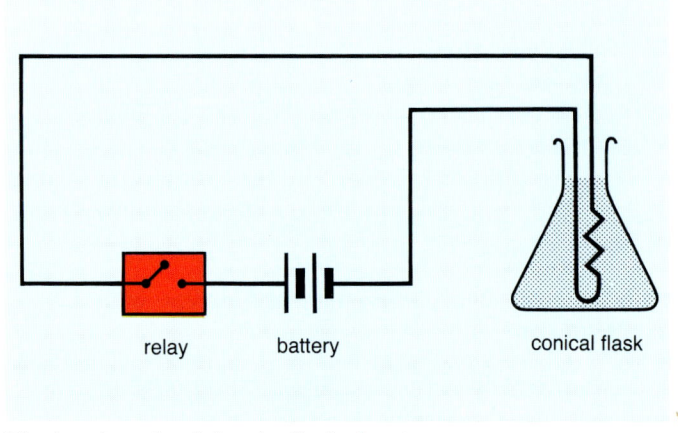

The heating circuit for the flask of water

44

IT AND MICROELECTRONICS

What you do

> **Safety!**
> **Do not get any water on the electric circuits!**

1 Connect the battery to the electronics kit and check that it is working.
2 The temperature sensor is *on* if the temperature of the water is *above* a certain level. Connect it to a NOT gate to reverse this signal. This means that it is *on* if the temperature is *below* the required level.
3 Connect the output from the NOT gate to an input of an AND gate. Then connect a slide switch to the other input.
4 Connect the output from the AND gate to the relay switch and a light.
5 Check that the circuit is working properly. The points below will help you.

- If the slide switch is off, the light should stay off whether the temperature is above or below the required level.

- If the slide switch is on and the temperature is below the required level, the light should be on and the relay switch closed.

- If the slide switch is on and the temperature is above the required level, the light should be off and the relay switch open.

You should be able to adjust the temperature at which the sensor switches on by turning a dial.

When the circuit is working properly, turn the slide switch off.

6 Connect up the heater circuit as shown in the photo and the diagram. Place the immersion heater and the temperature sensor in the flask of water. You can then put the thermometer or temperature probe in the water.
7 Make sure the temperature sensor is off. Turn the dial if you need to. Do not set it at too high a temperature.
8 Turn the slide switch on and start the timer. The light should come on, indicating that the heater is on.
9 Note down the temperature of the water at the start.

10 After a while, the temperature should reach the level at which the sensor switches on. This makes the heater switch off. Note down the time when this happens, and the temperature.
11 Continue your experiment until the heater switches on again. Note the time and temperature at which this happens.
12 Allow the control system to run a little while longer, until a pattern of switching on and off is set up. Record the time and temperature each time the heater switches on or off.

Recording the results

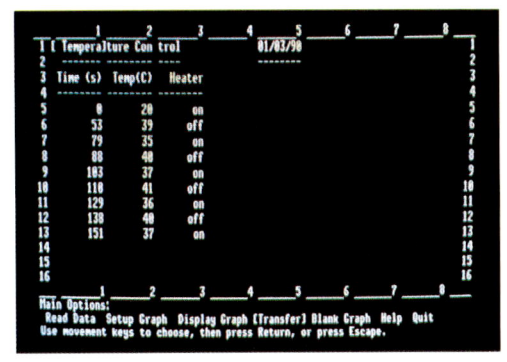

Your spreadsheet should look something like this

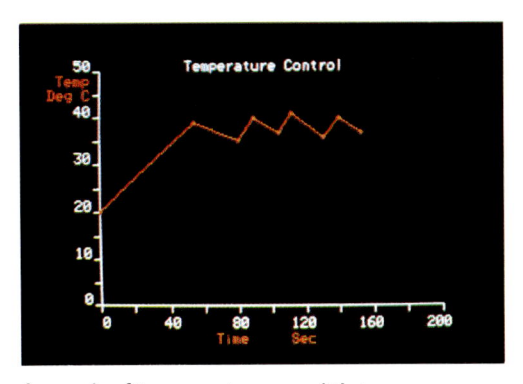

A graph of temperatures and times

13 Start your spreadsheet program running on the computer and open up a new spreadsheet.
14 Enter the start time (0) and temperature. You should then enter the times and temperatures at which the heater turned on or off.
15 Produce a line graph of these figures, plotting temperature (*y*-axis) against time (*x*-axis).
16 Print the spreadsheet and the graph. You can compare your results with other groups in your class.

45

ACTIVITIES

Questions

Look at your spreadsheet and graph.

a) At about what temperature does the heater switch off?

b) At about what temperature does the heater switch on?

c) Is the heater on for longer than it is off?

d) What is the range of temperatures of the water? (Ignore the time when the water is heating up at the start.)

e) Would a more powerful heater make this range smaller or larger? Try to explain your answer.

Taking it further

a) If you have a temperature probe which can be connected via a data logger into a computer, use this instead of the temperature probe, spreadsheet and graph display program.

b) (i) How could you keep tropical fish warm without using a logic circuit like the one in the last activity? Discuss with 2 or 3 friends how you could do this. What are the advantages of using an automatic system like the one in the activity? Can you think of any disadvantages?

(ii) Imagine that there is a power cut. This means that the fish tank heater will not work and you have to find some other way of keeping the water in the tank warm. In a group, discuss what you would do. Design an investigation to test how well your suggestion works. Check your plan with your teacher, and then carry out the investigation (you should try to use some electronic equipment in your investigation). Collect your data and compare your results with those of other groups in the class.

(iii) Logic circuits can be used to control many things, like the level of water in a tank or the direction of ships and aircraft. For each of the situations below, think of some other examples of this type of control:

at home; in a factory; on a farm; in a laboratory; in a hospital.

The automatic pilot on a passenger aircraft works in a similar way to the temperature control system in this activity

Glossary

Automatic control   Controlling a system by signals which are produced within the system itself.
Analogue signal   A signal which is proportional to the quantity it measures.
Analogue-to-digital converter (ADC)   A device which converts analogue signals into digital signals.
Computer   A digital, electronic device which processes information.
Communication   Moving information from one place to another.
Control   Making a system operate in a certain way.
Digital logic   Rules for combining digital signals to produce other digital signals.
Digital signal   A signal which is either on or off.
Electrocardiograph (ECG)   An instrument for measuring the way a person's heart is beating.
Electronic   Describes systems such as computers which use electricity but have no moving parts.
Frequency   The number of times something happens.
Histogram   A bar chart of the number of times things happen.
Information technology (IT)   The combination of computers, control systems and communications to provide useful services.
Input   Anything which goes into a system.
Mark:space ratio   The ratio of the time a digital signal is high to the time it is low.
Monitor   To measure something continuously.
Output   Anything which comes out of a system.
Probe   A measuring device.
Relay switch   A switch which is controlled by another switch.
Reset   To make a counter go back to zero.
Sensor   A measuring device.
Spreadsheet   A table of numbers and labels kept on a computer.
Switch   A device which turns an electric circuit on or off.
System   A collection of parts working together to carry out useful tasks.

Equipment

A list of the minimum equipment needed in this book is below.

Electronics kit

- **Input sensors**: temperature; light; press and slide switches on wires which plug into logic unit; digital thermometer, or analogue thermometer with analogue-to-digital converter.
- **Logic unit**: AND, OR and NOT gates; digital counter.
- **Outputs**: light; buzzer; relay switch; LED display.

Computer

General purpose microcomputer with spreadsheet and graph display software. If the activities are to be done in a 'circus', with 5 groups of 5 pupils, 3 microcomputers will suffice for a class.

Data logger

Optionally, a general purpose data logger with a variety of input sensors, able to transfer recorded data to computer, running software which stores and displays the data.

Laboratory and general equipment

Standard laboratory items are listed under '**You will need**' in each activity.

Acknowledgements

We are grateful to the following companies, institutions and individuals who have given permission to reproduce photographs in this book.

Allsport/Bob Martin (16, middle); Dr Alan Beaumont (39); The J Allan Cash Photolibrary (14, top left); Ford Motor Company Ltd (1, bottom); Griffin and George (11, top right); International Stock Exchange Photo Library (13; 18, middle; 27, two pictures); ITN (3, top); Metropolitan Police (9, bottom right); National Rivers Authority, Thames Region (11, upper left); National Westminster Bank plc (5, upper left); Roddy Paine (17, bottom; 20; 22; 24, bottom left; 28, five lower pictures; 34; 35; 36; 37, two pictures; 38, right; 41, top left and bottom left; 42; 44, bottom; 45, two pictures; 46); Racal Marine and Energy Publicity Department (32, left); Sainsburys plc (6, top left); Science Photo Library (6, bottom left; 7, top; 8, bottom; 10, two pictures; 11, bottom right and middle left; 14, middle; 15, top; 24, top right; 26, top; 32, right; 33); Swan Photographic Agency Ltd (5, upper right; 38, left); Colin Taylor Productions (1, top; 3, middle; 4, two pictures; 5, upper left, bottom left and bottom right; 6, three pictures upper right, middle left, lower middle and right; 7, four pictures bottom right; 8, top three pictures; 9, top left, top right, middle, bottom left; 11, middle right and bottom left; 12, three pictures; 14, bottom right; 15, bottom three pictures; 16, left and right; 17, top; 18, bottom left; 19, four pictures; 21, two pictures; 24, top left and middle left; 26, left; 28, top; 40; 41, right; 44, top); Telefocus, a British Telecom Picture (26, bottom right).

British Library Cataloguing in Publication Data

Bishop, Peter *1949–*
 IT and microelectronics. – (Science scene).
 1. Information systems. Technological development
 I. Title II. Series
 302.2

 ISBN 0 340 53270 X

First published 1991

Second impression 1992

© 1991 Peter Bishop

All rights reserved. No part of this publication may be reproduced or transmitted in any form or by any means, electronic or mechanical, including photocopy, recording, or any information storage and retrieval system, without permission in writing from the publisher or under licence from the Copyright Licensing Agency Limited. Further details of such licences (for reprographic reproduction) may be obtained from the Copyright Licensing Agency Limited, of 90 Tottenham Court Road, London W1P 9HE.

Typeset and illustrated by Gecko Ltd, Bicester, Oxon.
Printed in Hong Kong for the educational publishing division of Hodder and Stoughton Ltd, Mill Road, Dunton Green, Sevenoaks, Kent by Colorcraft Ltd.